FRACTURED CITIES

In both the United States and Britain, cities are undergoing changes which have intensified the divisions between regions, social classes and racial groups. These cities urgently need to develop new social, economic and political structures to meet both contemporary and future challenges.

Concerned with urban change and the workings of the capitalist market, *Fractured Cities* describes how public choice and regulation theory addressed the market, both during and after the Reagan and Thatcher years, and how the theoretical and practical policy issues created by the 'conservative revolutions' dramatically altered the political economy of the urban environment.

This book looks at the two-sided nature of urban change in relation to broad social, political and economic developments. The expansive aspects of the market economy and its tendency to move into recession have affected the polarisation of urban societies and the degree to which community groups gain access to political power. *Fractured Cities* assesses the implications of social disorder and the attempts of both the public and private sectors to offset economic decline and social conflict.

The book concludes with an assessment of emerging structures and approaches designed to extend local participation. These are viewed within the context of environmental problems, the concept of 'limits to growth' and the possibilities of different forms of political action.

Brian Jacobs is Senior Lecturer in Public Administration in the Department of International Relations and Politics at Staffordshire Polytechnic, England.

FRACTURED CITIES

Capitalism, community and empowerment
in Britain and America

Brian D. Jacobs

London and New York

First published 1992
by Routledge
11 New Fetter Lane, London EC4P 4EE

Transferred to Digital Printing 2004

Simultaneously published in the USA and Canada
by Routledge
a division of Routledge, Chapman and Hall, Inc.
29 West 35th Street, New York, NY 10001

© 1992 Brian Jacobs

Phototypeset by Intype, London

British Library Cataloguing in Publication Data

A catalogue reference for this book is available from the British Library

ISBN 0–415–07852–0 (HB)
ISBN 0–415–07853–9 (PB)

Library of Congress Cataloging in Publication Data
Jacobs, Brian D. (Brian David), 1950–
    Fractured cities : capitalism, community and empowerment in Britain and
America / Brian D. Jacobs.
        p.   cm.
    Includes bibliographical references (p. 264) and index.
    ISBN 0–415–07852–0 (HB). — ISBN 0–415–07853–9 (PB)
    1. Urban policy—United States. 2. Urban policy—Great Britain.
3. Urbanization—United States. 4. Urbanization—Great Britain.
5. United States—Economic conditions—1981– 6. Great Britain—
Economic conditions—1945– 7. United States—Social conditions 1980–
8. Great Britain—Social conditions—1945–
I. Title.
HT121.J33   1992
307.76′0973—dc20   92–294
                                                             CIP

CONTENTS

CONTENTS

FIGURES

TABLES

PREFACE

My thanks go to the numerous individuals and representatives of public and private sector organisations in the United States and Britain for the help which they have given during the course of this research. There are too many to mention individually, but some of the organisations which have made significant contributions are noted in the relevant chapters in the book.

This book was written following Department of Education and Science is Industrial Secondment to Business in the Community (BITC). The views in the book reflect those of the author and do not reflect those of BITC or any other organisation.

In addition to those organisations acknowledged in the footnotes to tables and figures, the following have kindly granted copyright permission for the use of material in the text and appended to tables: Central Statistical Office (London), Cox and Murray Inc., Gower Publishing Group, National League of Cities (USA), Random House Inc.

Brian D. Jacobs
July 1991

INTRODUCTION:
URBAN CRISIS AND CHANGE

This book is about urban change and crisis in the United Kingdom and the United States of America with particular reference to the 'conservative years' of the 1980s and early 1990s. A comparative analysis should cover specific national characteristics in terms of governmental organisation, political cultures, attitudes and economic change (Eldredge, 1967; Page and Goldsmith, 1987; Norton, 1990).

WHY THE UK AND THE USA?

This book raises questions which may be relevant in more comprehensive multi-nation comparative studies. These particularly relate to the international context of national problems and the economic implications of economic restructuring and urban change. The book examines two nations which have faced often similar social and economic problems. Both countries are industrialised 'urban societies'. In both nations, the cities have suffered the consequences of economic decline and acute social distress. There has been an increasing internationalisation of their economies which has directly affected both urban growth and decline. British and American financial markets are now part of the global system of exchange and British and American companies in financial services, property, commerce and industry invest and exchange personnel.

The political systems of the two nations differ, although there was an ideological affinity between recent UK and US conservative governments. Prime Minister Thatcher came to power in 1979. President Reagan was elected in 1980. They and their respective successors, John Major (who became prime minister in 1990) and George Bush (who was sworn in as president in 1989), were associated with a trend in favour of a more fundamentalist conservative politics which placed greater emphasis on the role of private initiative and called for effective controls on public expenditure. There are also other notable factors in common:

1

INTRODUCTION

- Both countries are English-speaking.
- Both have advanced industrial economies, although there are differences in wealth: the US gross national product per capita in 1990 was estimated at US$21,850, while in the UK it was US$17,000.
- British settlers dominated early America so that a British identity became a strong influence on American thought about issues and relations between the two countries.
- American political and legal institutions derived from the British, yet the two systems developed widely different political and institutional arrangements.
- There are Protestant majorities in both countries – 50 per cent in Britain and 64 per cent in the USA.

(Davis, J., 1986)

Frequent reference is made in the media and elsewhere to patterns of life in the distressed districts of UK and US cities. Although there are important differences in demographic and social patterns, transatlantic cultural links are often popularly expressed in terms of the rise of black hip hop music, the fashion styles of 'the ghetto' and the common experiences of ethnic minorities in depressed urban environments. In such matters there may be said to exist a special and close affinity between British and American city cultures.

The Global Context

Despite the connections and differences between the UK and the USA, the present study is inevitably limited in scope as a result of its two nation focus. The further development of a comprehensive comparative perspective is likely to come from research into the international process of change (Clark, 1988). Future research will doubtless continue to view urban change in the advanced industrialised societies (Borchert et al., 1986; Cheshire and Hay, 1989; van Vliet and van Weesep, 1990) and the Third World (Thrift and Forbes, 1986; Phillips and Yeh, 1987; Henderson and Castells, 1987). It will also consider the problems of cities in the post-Communist countries of East Europe (Szelenyi, 1983; Hegedüs and Tosics, 1990; Mandič, 1990). In addition, there is scope for further research into the rapidly expanding regions and 'mega' cities of Asia and Central and South America (Hamilton and Harding, 1986; Kirkby, 1988; Hall, 1989). New patterns of urban growth and the implications of technological futures inherent in developments such as Adelaide's Multi Function Polis (MFP) and similar proposed centres of high technology and supercommunications underscore the need for coordinated international research initiatives.

2

A RESEARCH AGENDA

There are fundamental political, economic, social and globally significant considerations involved in this task. This assumes that the problems of the cities have to be seen as not narrowly defined spatially. An approach is required which goes beyond the development of a focus on particular districts or communities. It demands more than an examination of problems set within given local government boundaries. The 'urban' politics of 'place' is not rendered redundant as such, but set within a broader frame. The tendency to view cities in their wider setting is evident in pluralist approaches, regime theory, market-centred theories of urban change and recent class analyses (Waste, 1990).

Comparative Studies

The present study covers a range of themes concerning political economy and organisation in a complex and ever changing environment (Øyen, 1990). Heidenheimer *et al.* (1990) have recognised the value of comparative public policy in revealing the nature of the complexities involved. They refer to the different approaches that have been used to develop comparative studies, starting with *socioeconomic theories*. These concentrate upon the processes of economic growth and how nations respond to them. Economic growth affects levels of expenditure on welfare programmes and patterns of economic and social interaction between rural and urban areas.

Another approach looks at *cultural values* and the consequences of different approaches to economic management arising from different traditions of statist paternalism or laissez-faire liberalism. A *party-government* framework develops the theme that 'policy matters' when comparing nations. This was often argued when researchers attempted to show that political party commitments made a difference to welfare outcomes.

A further approach concentrates on the importance of *political class struggle* where the emphasis is upon the contest between business and labour under capitalism. *Neo-corporatist* approaches develop the idea of a political accommodation of interests under capitalism. The emphasis in these studies is on the connections between governments, interest groups and business, and their coordination at the national and local levels. *Corporatism* expresses the trend towards the obliteration of the boundaries between public and private sector interests (Cawson, 1985).

Heidenheimer *et al.* also allude to the *institutional-political process* perspective which puts the state at the centre of the stage. This involves an analysis of the historical development of state institutions in different countries, the ways in which institutions interact and the feedback effects on policy.

For Heidenheimer *et al.* (1990, p.9) these approaches are not mutually exclusive or exhaustive. Consequently, there has been a move away from 'single-factor deterministic theories' towards a recognition of the need to integrate diverse perspectives to produce plausible accounts of policy development. This has been made difficult by the problem of finding truly comparable data; problems with developing comparative policy performance indicators; disagreements over definitions of, for example, concepts of welfare and equity; problems in comparing the relationships between variables (for example, there may not be enough international data available to assert definite relationships between assumed causes and effects).

One partial solution is to develop broad-scale comparative mapping of similarities and differences and to produce contextually rich individual case studies (Heidenheimer *et al.*, 1990, p.12). Comparisons can be made without deciding in the abstract whether it is the similarities or differences that matter the most. Guidelines may be established for identifying certain relationships and historical lessons without rendering the whole exercise invalid. In spite of the tentative nature of some of the conclusions, there are important insights to be gained by looking at the following:

1 *Comparisons of economic and social changes* and the ways in which domestic and international market forces impact upon urban change.
2 *Public attitudes* with reference to modes of political organisation, interest group demands and *changing political cultures*.
3 *Institutional/political differences* in terms of conceptions of government, the relationship between central and local government authorities and their responses to external economic and social change.
4 The *changing nature and scope of public policy* which needs to view 'urban issues' in global terms, particularly as policy relates to environmental questions and the future quality of life.

The present work therefore draws upon various theoretical contributions which have added to our knowledge of 'urban politics' and comparative public policy. While there are limitations with a multi-theoretic focus, it is recognised that 'complete' theories (such as regulation theory and public choice) have their own problems. They have often provided useful overviews of the processes of change and market behaviour, but empirical problems and inconsistencies inherent in such theories have reduced their value. It is thus mistaken to assume that there is any current single paradigm which adequately theorises about urban change.

Dunleavy's (1991) critical view of public choice seems to provide one useful avenue for the development of a multi-theoretical analytical framework. He accepts the value of rational choice but separates it from its present largely 'new right' or 'conservative' underpinnings. This opens the way for a re-examination of rational choice models in conjunction with the insights provided by regulation theory and theories of political accommodation.

The objective here is not to provide an alternative overarching urban political economy, but to offer a formative perspective in the spirit of Dunleavy's search for a new synthesis, and one which is also specifically concerned with urban change and crisis. The book therefore defines a series of issues which constitutes an agenda for comparative urban research. It points to a range of important trends and policy questions which will require fuller examination, especially as researchers become more concerned with global issues.

THE 'URBAN' DIMENSION

Global concerns imply that 'urban' politics can not be defined in terms of a narrow spatially defined field. Harvey (1988, p.304) argues that 'space' is important since spatial forms incorporate social processes. However, for Harvey, 'urbanism' is not simply 'a thing' in the ordinary sense of the word. A city can be regarded as a set of 'objects' arranged according to some pattern in space, but that view of cities is far too simplistic. Urbanism involves a set of social relationships which reflects the relationships established throughout society as a whole. A conflict exists between social processes and what Harvey calls 'the static geometry of the spatial form'. There is a social-process–spatial-form interaction. The distinctive role which space plays in both the organisation of production and patterning of social relationships is 'consequently expressed in urban structure' (Harvey, 1988, p.307).

Urban structure is not merely fashioned out of spatial logic. For Harvey, it reflects ideologies (such as the urban versus the rural) and has a 'certain autonomous function' in fashioning the way of life of a people. Urban structure affects social relationships and the organisation of production as well as itself being 'channelled and constrained' by forces and influences which are fundamental to the economic base of society. The city as a built form and 'urbanism' as a way of life thus have to be considered with reference to their interrelationships. City, suburb and rural areas, are therefore best viewed as incorporated within the wider 'urban process' (Harvey, 1988, p.307).

The scale and diversity of city organisation within the urban process has produced a variety of problems and conflicts which indicate the makings of crisis and the fracturing of the urban social, political and economic fabric. There are conflicts between rural and urbanised areas, central city and suburb and metropolitan centres in the western world and cities in underdeveloped nations. There are conflicts between races and classes and public and private interests. Harvey argues that these conflicts, in the west, are related to the restructuring of cities in response to capital's desire to accumulate and invest. Capital finds its main markets in urban economies and responds and adapts to new needs and effective demands. Urban change is thus an imperative for capital as part of the process of accumulation but it also poses many problems.

MAJOR THEMES: CRISIS AND CHANGE

Change and its related tensions has long been the focus of research concerning transformation and strain in urban systems (Sert, 1942; President's Commission, 1960; Banfield, 1971; Sternlieb and Burchett, 1973; Lawless, 1981; Cardew, Langdale and Rich, 1982). During the 1960s and 1970s, urban problems were regarded as being of national significance following the breakdown of social order in America's major cities. Watts, Harlem, Detroit and Newark were places that went down in history as representative of the anger of the ghettos. Black Americans expressed their resentments against the white 'power structure' openly on the streets (Boutelle, 1967) as the use of the National Guard and federal troops with tanks in domestic riot situations dramatically drew attention to their problems. This raised serious questions about the ability of public authorities to deal with the consequences of economic decline and social discontent.

Uncertainty and the problems of debt and inflation meant that crisis was an integral part of the process of change in the British and American market economies into the 1990s. 'Crisis' exists where there are economic conditions which produce serious conflicts in society. These deprive subjects of the full realisation of their rights and social opportunities and the assurance of political and economic stability (Habermas, 1973). Crisis situations often concern conflicts over scarce resources or situations where there is mounting anxiety about the plight of a group or groups in society. Crisis situations imply a lack of confidence in the future or a decline in confidence to levels below those previously enjoyed.

British and American cities suffered in this way. They were affected by moves away from Keynesian interventionism by the public sector in fields such as education, housing and urban renewal. In the USA, Reagan adopted a negative policy towards urban renewal and his administration initiated the termination or gradual elimination of many community and regional development programmes. This reflected a belief that the promotion of community and regional development was not an appropriate activity for federal government and that such a role should be assumed by the states, local government and the private sector. The administration regarded many urban programmes as being inefficient and better administered by local administration which was said to be in a sounder financial condition following the managerial changes made during the 1980s.

INTRODUCTION

Urban Britain

In both Britain and the USA, the cities struggled to generate economic resources to stimulate urban renewal and the improvement of the urban environment (Sills, Taylor and Golding, 1988). The problems of the cities did not start in 1979 or 1980. A legacy of decline produced problems which were historically induced by conditions which were often beyond the power of governments to control (Rees and Lambert, 1985; Gamble, 1985).

This view was implicit in Michael Heseltine's (1987) analysis. Heseltine was a leading Conservative party leadership contender in 1990 and Secretary of State for the Environment in both the Thatcher and Major governments. Heseltine pointed to the passing of Britain's formally dynamic phase of capitalist development. He referred to the distorting effects of the market which had shifted resources away from provincial cities towards London and the south-east. He produced a radical analysis of capitalist financial institutions and urban decline by calling for measures to divert resources into productive investments and infrastructure. This would be an effective way of overcoming the problems faced by the 'cities in crisis' (Heseltine, 1987, pp.151–76). He also advocated the adoption of US-style elected executive mayors to replace the British 'amateur' council leader system.

The British urban riots of 1981 and 1985 forced consideration of the impact of economic policy upon communities. The Heseltine analysis served as an important reminder that, even within the Conservative party, there was concern about the efficacy of market-led policies and their impact upon social cohesion and harmonious community relations. The problems were to grow in the 1990s as Britain continued to suffer from long-standing economic problems and social conflict (Overbeek, 1990).

Urban America

In the USA, social indicators showed a worsening of the condition of distressed urban communities in both social and economic terms. Urban tensions in America were a problem, with evidence showing that racial conflicts, problems with drugs and gang warfare were getting worse. Urban America was still prone to riots as the outbreak of Hispanic disturbances in Washington DC in May 1991 clearly illustrated.

The decline of major sectors, such as the automobile and steel industries, contributed to local distress and increases in unemployment even among skilled workers. Many of America's cities were physically run-down, socially demoralised and environmentally polluted (Gottdiener, 1987). Welfare roles continued to expand as social programmes were hard-pressed to cater for the demands of those affected by adversity.

THE ELEMENTS OF CRISIS

Given such developments, the general crisis affecting the cities and the debate about urban issues may be viewed in terms of at least five components.

1 *An economic crisis* which involved the decline of manufacturing in traditionally 'industrial cities' and the growth of unemployment. The economies of the USA and the UK were facing increasing competition from abroad and mounting levels of inward foreign investment. Economic problems affected city governments through the reduction in their tax bases, inflation of the costs of services and the burden and cost of infrastructural decay.

2 *A social crisis* which produced a polarisation between the 'underclass' and the 'better-off' in society. *The social class divide* in British and American cities arguably exacerbated communal tensions between different racial and ethnic groups as crime, racism, drugs and domestic violence increased. It also placed a heavy cost burden on national and local governments faced with mounting bills for social provisions and benefit payments.

3 *A political crisis* concerning representation and *the political incorporation of community leaders*. Governments promoted partnership in urban renewal. However, in the USA, there was mounting disillusion about the inability of community leaders to achieve beneficial economic and social changes. Some prominent black political leaders were discredited in the eyes of their supporters following disclosures about their personal lives and political activities. Militant organisations grew in strength in many communities.

4 *A crisis of ideology.* Welfare-liberals and interventionist socialists failed adequately to address the ideological challenge mounted by the right on the major social, political and economic changes taking place in the 1980s and 1990s. The most 'radical' and coherent 'alternatives' tended to be produced by those on 'the new right'.

The early 1990s saw the re-emergence of economic recession in both the UK and the USA. In Britain, the government expressed support for 'the social market' while John Major advocated a 'classless society' based on merit. Bush advocated the market approach, but continued to expand certain domestic programmes. The right, both in practice and in theory, was faced with coming to terms with the implications of expanding welfare expenditures connected with unemployment and recession. New right prescriptions for free market solutions were thus contradicted by the performance of conservative governments which seemed to be unwilling or unable fundamentally to reduce state expenditures.

5 *A global crisis of the cities* concerning the role of cities in the context of the world debate about the future of the planet and its ecosystem.

8

One key question was how long could the big cities keep expanding without incurring serious environmental dislocation?

ECONOMY AND IDEOLOGY

In the 1980s, new right economists, such as Milton Friedman, stressed the importance of the market. In the USA, the 'new economics' rested on three basic assumptions. First, that sustained economic growth was essential to generate wealth throughout society. Second, that the economy was prevented from growing because of unfilled investment needs caused by insufficient capital formation. Third, that this insufficiency should be remedied by tax reductions which would stimulate saving which in turn would provide resources for investment. This would assume that there was also a reduction in economically inefficient and inflationary social expenditures which had a depressing effect upon employment and overall demand in the economy (Eckstein, 1981).

Such thinking influenced Reagan and Thatcher. Reagan, although sharing much of Thatcher's basic philosophy, was faced with different conditions. In the USA, several sectors of the economy were poised for rapid expansion (computers and electronics for example) while many cities enjoyed boom conditions (particularly in the so-called 'sunbelt'). Reagan's policies required substantial spending on defence which made it difficult to keep down public spending unless there were dramatic cuts in domestic programmes. Reagan therefore placed the blame for an increasing budget deficit on Congress while favouring substantial reductions in expenditure on social programmes. Deficit financing contributed to economic expansion in the short term, although by the end of 1986, the price of this expansion was realised in terms of a slowing of the economy and the initiation of further policies to reduce the public debt burden.

Underlying Reagan's economic strategy, was the distinctly Thatcherite philosophy that social programmes should be constrained, subsidies be reduced and a general reduction in tax burdens for all sections of society be achieved. In this way, the poor would be encouraged to make provisions for their own futures and take advantage of opportunities in 'the marketplace'. Companies and non-profit organisations were expected to contribute more in supporting urban renewal through initiatives designed to promote local enterprise and employment.

Changing Attitudes: Rolling Back Socialism

Such developments were accompanied by attempts to foster attitudes conducive to the development of enterprise and to the strengthening of values conducive to the consolidation and expansion of capitalism. Thatcher called for the development of the 'enterprise culture' and for the elimination of socialism as a major political force. Drawing on American experience, she passionately advocated the attributes which she so admired about the American spirit of entrepreneurialism. The promotion of the capitalist ethic and the return to what she regarded as 'Victorian values' was a crucially important aspect of the development of her new urban agenda.

Community organisations, tenants' groups, the young and the unemployed were all to be encouraged to identify with capitalist initiative and the benefits of the market. The deprived communities of the cities were urged to associate with the private sector through partnerships adhering to a new 'shared vision' of urban renaissance.

Both the Thatcher and Major governments represented an alternative to socialist interventionism. Their challenge was part of an attempt to produce wealth by encouraging the inter-play of market forces, sound monetary policy and prudent management of public borrowing. If this meant that inefficient companies would have to 'shake-out' labour, then it was a necessary part of the process of economic restructuring. The public sector would also have to 'shake-out' to reduce its own inefficiencies and redefine its expectations about what services it should provide.

POLITICAL CONSTRAINTS

Despite the challenge to the left's traditional commitments, it is important to take account of the *institutional and political constraints* faced in implementing radical conservative programmes.

Rolling Back Thatcherism?

Thatcher enjoyed the advantage of having a considerable degree of executive control over legislation. Despite this, her dramatic removal from office in November 1990 indicated that the pressures brought to bear upon a British prime minister could result in the deposition of even the most strident and confident of politicians. In the case of Thatcher, her own party expressed worries about her style of government and the unpopularity of policies, especially the deeply resented Community Charge ('poll tax'), with which she had been personally identified in the minds of the electorate.

John Major's electoral popularity was initially linked to the fate of the tax and his ability to deal with the complex problem of its replacement. The Labour party and Liberal Democrats exploited the unpopularity of the poll tax and the government's administrative wrangles with local authorities which resulted from attempts to reduce the direct burden of local taxation before the coming general election.

Checking the 'Conservative Revolution'

In the USA – both Reagan and Bush found it difficult to gain support for measures which were branded as ideologically motivated. In the 1981–82 Congress, Reagan was helped by the mood of a conservatively minded House of Representatives and Senate. This was despite the fact that the House was under Democratic control. Congress was influenced by the shift in public opinion to the right, and there was a strong representation of conservative southern 'boll weevil' Democrats who frequently voted with the Republicans on social and economic issues. This created a conservative 'coalition' which allowed Reagan to press ahead with radical changes to domestic programmes.

In the Congressional and gubernatorial elections of 1982 there was a swing away from the Republicans just as the 'Reagan revolution' seemed to be at its height. The shift was strongest in the depressed industrial areas of the Midwest and in the south. The Republicans managed to retain their Senate majority, but suffered large losses in the House. Many of the losers were radical conservatives who had entered Congress on the crest of the conservative wave in 1980. The elections of 1982 disrupted the Congressional conservative majority making it more difficult for Reagan to carry out social reforms. This was particularly evident during 1983 and 1984 when Reagan increasingly resorted to his power of veto on major issues.

In the 1984 Presidential election Reagan was again returned. The Democrats were divided between conservatives and a beleaguered liberal wing opposed to the Reagan roll-back of welfarism. They were also hampered by their lacklustre presidential candidate, Walter Mondale.

During 1985, Reagan faced opposition from House Democrats over his budget package and tax reform measures. Presidential policy initiatives were increasingly less attractive to Congress. Derbyshire (1987) suggests that 1985–86 marked the end of the 'Reagan revolution' as there was a sharp drop in Congressional support for the President both on domestic and foreign affairs issues. This was underscored when the Democrats gained control of the Senate in the 1986 mid-term elections with a fifty-five to forty-five seat majority margin. The President was further weakened by the Iran–Contragate scandal despite a renewed attempt at consensus-building in Congress spearheaded by Howard Baker as new Chief of Staff.

George Bush won the 1988 Presidential election for the Republicans standing against Democrat Michael Dukakis. However, Bush was unable to bring additional strength to the Republicans in Congress in elections in the same year. The Republicans lost three House seats despite the evident public desire to play for safety with Reagan's former vice president.

Bush, facing a potentially hostile Congress, adopted a conciliatory position. His policies represented a certain continuity with Reagan-style conservatism. This did not, however, imply that he sought to apply a rigid 'game plan' or that he was in any way 'Reagan's man'. Bush imprinted a conservative but pragmatic mark on policy. Indeed, the right accepted that he would loose sight of some of the conservative principles acknowledged by Reagan.

Beckworth (1988) argues that some Democrats in Congress saw this as an opportunity to assert their own agenda on a range of issues and that the prospect of compromise enhanced their chances of success in influencing important policy initiatives. This was consistent with Bush's Presidential campaign commitments which envisaged the allocation of extra funds to the states for educational provisions, an increase in funding for the Head Start programme for preschool children, the promise to create jobs and increase subsidies for child daycare. It was such commitments which threatened some of the former Reaganite assumptions about the evils of federal handouts to the poor and alerted free marketeers to the dangers of an assertive federal bureaucracy.

The pressure to compromise with Congress was evident during the 1990 controversy over the US budget. The President was forced to accept the passage of tax increases as part of a package of measures designed to control the federal budget and reduce a massive budget deficit. The Democratic Congress asserted its power by attempting to mould the budget. In so doing, Bush appeared to be in a relatively weak position following the breakdown of party coherence in Congress and a steep fall in his public opinion ratings. However, the 1991 Gulf War victory over Iraq transformed the political climate and significantly strengthened Bush's position. The political benefits of victory ensured that the President's popularity in the opinion polls rose dramatically. Democrats were swept along with the triumphal tide as Bush drew a connection between the speedy Allied victory in the Gulf and the need to hasten domestic legislation through Congress.

GOVERNMENT, POLITICS AND SOCIAL STRESS

Despite the initial conquests of Thatcher's and Reagan's conservative 'revolutions', they did not achieve the kind of radical changes expected in the early 1980s. Reagan's policies accompanied an increase in the national debt. Defence spending increased in the early 1980s, and there were compromises on the original commitment to reductions in domestic programme expenditures. Federal spending did fall as a percentage of national income, but only slightly. State and local government spending increased (see Chapter 5). The Bush administration continued to spend on new defence initiatives and supported substantial public intervention in parts of the economy such as in the semiconductor industry through the Sematech consortium.

In Britain, the Thatcher government presided over higher overall government outlays even after adjustments for inflation. This was achieved with only marginal reductions in public spending when expressed in terms of its share of gross domestic product. The Major government increased public sector borrowing and sanctioned additional central funding for local government to offset the affects of the poll tax. Major also envisaged an upward percentage share of GDP on public spending.

Resilient Local Politics

Public programmes at the local level survived despite expenditure constraints and privatisation. Private initiative often provided enhanced, not reduced, opportunities for local-level interest groups to benefit from business development programmes and other activities. Community groups could be accommodated 'within the system' despite the doubts of many activists about the virtues of 'the market'.

Domestic programmes suffered, and policy priorities changed within the overall pattern of public spending. However, the need to maintain employment levels and social order called for government expenditures. In urban policy, neither the Thatcher or Reagan administrations could simply entirely 'withdraw' central funding from the cities. Heseltine's plea that the Conservative party should not 'retreat from the cities' was thus apposite. In Britain, political considerations played an important part in persuading the Thatcher government that determined policies were preferable to inaction. The establishment of urban development corporations, inner-city task forces and expanded training schemes for the unemployed were indicative of interventionist public policy initiatives.

In the USA, the structure of the US tax regime, and the ability of local governments to raise money through the issue of tax-exempt bonds maintained public involvement despite federal constraints. State government funding of programmes which had lost federal support, and the central policy promoting private initiative in local programmes, served to highlight the continued importance of a public role. The intervention of state and local governments in partnerships with the private sector illustrated the vital role of public agencies in promoting economic growth.

Political Incorporation

All this implied cooperation between government and community. During the 1960s, urban politics was dominated by issues concerning access to public and private agencies and the 'powerlessness' of urban communities (Katznelson, 1981). In Britain and the USA, there was support for programmes designed to overcome social tensions, avert riots and promote local economic regeneration (Edwards and Batley, 1978). However, by the end of the 1970s, the prospect of dramatic change through government action had faded and the problems remained.

Community leaders gained access to local centres of power. In the USA, blacks scored notable successes in expanding minority representation, although by the mid-1980s many complained about the lack of economic progress despite the formal realisation of civil rights. Other minorities, such as the Hispanics, were faced with persistent economic disadvantage. Their position in the political process was weaker than black Americans, and their political status in society tended to be generally lower. This was reinforced by US government controls on immigration and the repatriation of illegal immigrants. As in Britain, repatriation and immigration control affected domestic political issues. Racism placed a stigma upon minorities which prevented them from advancing politically and restricted their economic opportunities.

PERSPECTIVES ON CHANGE

Political incorporation, the role of interest groups, governments and the private sector are examined in the following chapters within a perspective which assesses contesting theoretical approaches to questions concerning the economy and markets, social polarisation and the nature of urban change under capitalism.

Chapter 1: Perspectives

The late 1970s and early 1980s saw the left adapting to the politics of enterprise and individualism in the wake of changes in public mood and the ideological onslaught of the new right. The conservative 'public choice' prescription was to 'free' the market by removing government regulation, promoting privatisation and encouraging the development of the private sector.

The right's prescriptions could have a popular appeal, which was a fact that profoundly disorientated the left by challenging its view of the role of the state and notions of public intervention to alleviate social distress. Even so, the recession of the 1990s created problems for both the left and right – both being unable to provide satisfactory responses to the problems associated with market failure. The right asserted its confidence in the market while the left pointed to the development of new forms of capitalist regulation and organisation as a response to crisis.

Chapter 2: The Role of the State

Market activity and technological and economic change need to be examined with reference to the role of the state. Public choice theory neglected the importance of government in economic development. It failed adequately to explain the complex relationship between the state and the market and the nature of the interests involved in economic policy-making. However, caution was required when considering alternative neo-Marxist interpretations of change. To imply that there was an historical inevitability about certain aspects of change under capitalism (such as 'post-Fordism') was little more helpful in understanding the role of government than the 'market determinism' in public choice theory.

Public choice theory explained change by viewing government as being relatively weak and reactive when faced with the market. However, there were market factors which worked in favour of governments even if the tide of events often seemed to run against them. Chapters 3 and 4 examine the nature of the market conditions affecting the cities and the reactive role of government. Chapter 5 reviews initiatives and organisational factors which tend to give governments a more proactive role.

Chapter 3: Urban America: High-Technology and Poverty

The post-Second World War era saw the gradual challenging of the world economic position of the USA. In the 1980s, the USA faced increasing competition in international markets which caused domestic concern about the incursion of foreign companies. The continued internationalisation and structural change in the domestic economy played an important part in undermining the traditional economic foundations of urban America.

Urban change, and the decline of traditional industries, revealed the imbalances resulting from capitalist restructuring. Population shifts involved people and industries locating in areas away from traditional city-central locations to suburban 'corridors' or 'sub-cities'. The older cities witnessed the decline of their staple industries and abandonment of declining central city areas. The competition between cities for investment in new and existing industries developed an international character.

Technological advances accelerated alterations in the economies of the cities. The shift to service employment had social, political and economic implications. *Social polarisation* between those who benefited from economic expansion and those who did not worsened.

Chapter 4: Urban Britain: Enterprise and the Underclass

In Britain, the economic problems of the cities were similarly linked to Britain's declining international competitiveness. As in the USA, one indication of the fracturing effects of economic decline was the destruction of large segments of the country's manufacturing base. City economies rested on a precarious economic balance between growth in some sectors (as in financial services and high technology) and decline in others. *Social*

15

polarisation was intensified by the stark contrast between the condition of the underclass and the prosperity of those who had benefited from the 'enterprise culture'.

Chapters 5 and 6: Local Government and Urban Policy in the USA and the UK

There were important *institutional and political differences* between the UK and the USA which affected urban policy. Local government was organisationally complex in the USA. Nevertheless, it was based on a commitment to local democracy and the involvement of citizens in the political process. This tended to strengthen local government against the incursions of the centre.

American local governments were under financial pressure during the 1980s and 1990s and had to battle hard for resources. In the UK, there was a substantial erosion of the powers of local government as British local authorities operated in a hostile political environment. This effected the continued retreat of the 'urban left' and intensified controversy about the future structure, funding and autonomy of local government.

Chapter 7: Incorporating Community Politics

Fiscal stress at the local level meant that interest groups were faced with a highly competitive situation as far as resources were concerned. Rational choice approaches to interest group politics have stressed the importance of the costs and benefits to groups and their members of mobilising for public funds. Groups try to maximise a range of 'benefits', both economic and political, and these influence group attitudes to government and their willingness or otherwise to cooperate with politicians and public officials. Incorporation results from a desire by group leaders to maximise resources.

During the 1980s, despite economic constraints, even quite militant groups saw the benefits of cooperation. Changes in the political culture affecting local politics were producing new patterns of political activity and effecting changes in attitude in favour of market-led urban renewal. Social class issues were redefined in this setting as partnership and consent replaced conflict and dissent in the vocabulary of many local activists.

Chapter 8: Minorities: 'Empowerment' and Dissent

The 'incorporation' of black politics in the USA did not always produce perceived benefits or significant community-level 'empowerment'. There were important lessons to be learned about this in Britain especially in matters concerning social disorder, crime, drugs and urban renewal. The quest for representation in the political system did not necessarily imply that ethnic minorities would immediately benefit from a closer association with the state or with the 'visions' of the private sector.

Chapter 9: Corporate Visions

Business leadership and the identification of 'community goals' were seen by conservatives as crucial ingredients in successful urban renewal in both the USA and Britain. This chapter explores the role of the private sector and discusses different types of public–private sector partnership covering experiences in major cities in both countries.

Partnerships generated new political and economic structures which operated outside traditional local government frameworks. They mobilised community interests around a 'shared vision' of the future in unison with business and enabled companies to 'cartelise' their development activities.

Chapter 10: Initiatives Beyond Charity

Initiatives involving non-profit organisations were regarded as providing attractive partnership opportunities. In Britain, charities assumed an expanded role. In the USA, the tax and legal regime favoured the growth of non-profit organisations. The expansion of non-profit organisations in the USA enabled non-governmental agencies to take advantage of incentives which facilitated wider participation.

Chapter 11: Wither the Cities?

Chapter 11 discusses the challenges posed by the worsening environmental condition of major cities and the importance of *global policy concerns*. The growth of the green movement and the expansion of locally owned and controlled enterprises exemplified the vitality of local activism. Communities can identify needs and achieve change without necessarily depending upon the visions of public bureaucracies or corporate partnerships, but policies need to be bold and imaginative to arrest the environmental deterioration likely in the wake of further urban expansion.

1

PERSPECTIVES

Gabriel Almond (1988, p.853) notes that pluralism and Marxism now represent 'waning paradigms of contemporary political science'. While this is probably an over-simplification, as Almond admits, both the pluralist and Marxist approaches were affected by the onslaught of 'new right' thinking and both initiated important theoretical reassessments which had important implications for urban analysis. This chapter considers the renewed interest in conservative and neo-liberal approaches and provides a critical examination of public choice, collectivist and pluralist approaches to urban political analysis.

A NEW CLIMATE

Theoretical problems for the left were associated with real-life political change. The British Labour party underwent a major review of its policies and took steps to reduce the influence of its left wing. Prominent Labour politicians such as Brian Gould advocated market solutions as the way forward for Labour (Gould, 1989) and argued that the socially responsive 'socialist market' would be the answer to attacks against the welfare state (Miller and Estrin, 1986; Goss, 1988; Stoker, 1988).

In the USA, the new conservative agenda affected some of the most out-spoken politicians. The 1988 Presidential bid by Jesse Jackson was initially widely regarded as being representative of the interests of the urban poor and more in line with the radical campaigning traditions of some of the former liberals who had taken up the demands of the deprived. Jackson's desire to mobilise blacks and other minorities struck a populist chord in the ghettos and even attracted support amongst significant sections of the white electorate. His attempt to consolidate a 'rainbow coalition' which would give minorities greater influence in government was a clarion call to those who believed that a radical set of policies was necessary to improve the condition of the poor and open American government to the effective representation of their interests. For Jackson, this was the best way to pressure government to grant economic reforms and to campaign against the social problems of the cities.

Jackson's decision to support the cautious Democratic Presidential candidate, Michael Dukakis, signified his willingness to make compromises with a 'technocrat' who was supportive of only a modest programme of policy initiatives to aid the cities and the poor. Even though the Jackson campaign had pointed to the need for changes which would address the problems caused by market failure, there was still an assumption amongst liberals in the Democratic party that the market would have to be the main motive force to provide new opportunities for communities.

Walker (1990) assesses the impact of political change in the USA in terms of a move away from traditional working class activism and the 'gentrification' of the left. This involved the left reducing its links with the trade unions and becoming less reliant on Marxist-style 'vanguardism'. This was marked by a reorientation in response to changes in the structure of the labour market and the development of 'issue politics' around concerns such as the environment, Aids, race and gender. Many of the small groups active in these areas orientated towards the Democratic party.

THE 'NEW RIGHT'

In the 1980s, the appeal of 'the market' implied that individual rights would increasingly be linked to the democratising effects produced by market activity. This was seen by the right as an essential aspect of the liberation of the individual from collectivism and socialism. Much of the inspiration for this reaction against welfare-liberalism and state socialism came from the 'new right' (King, 1987a). Despite the radical aspect of this attack, The new right's analyses is often described as consisting largely of reformulations of 'traditional' conservative and liberal ideas (Levitas, 1986b).

Levitas (1986b, 1986c) argues that there was considerable variance between those who described themselves as new right. Writers, such as Bosanquet (1983), used the term to refer to a form of neo-liberal laissez-faireism influenced by Schumpeter, Hayek, Milton Friedman and others. Green (1987) traces the roots of the new right to the economic thinking of the Chicago School of economists such as Milton Friedman, to Friedrich Hayek and the Austrian School, and to public choice theorists.

For ease of exposition, the definition of the new right used by Ashford and Davies (1991) is adopted here. This refers to the new right as the entire collection of conservative and neo-liberal movements which have grown up in North America and Europe since the 1960s. The new right is not an ideologically homogeneous group, but consists of a variety of trends, many of which do not fit into the traditional left–right spectrum. For Ashford and Davies, the primacy of the debate about economics has tended to highlight the common ground between those on the new right and has obscured the 'many and profound' divisions over a range of political issues. This applies even within the neo-liberal and neo-conservative sub-categories where there are several separate groups, schools or movements.

Ashford and Davies point to the traditional conservative movement in Britain and the USA, represented by figures such as Scruton (1984) and Kirk (1953). Conservatives of this kind should not be confused with the American neo-conservatives such as Kristol (1983) and Podhoretz (1979) who combine a strong social commitment with a market orientation. In turn, this group is different in orientation to the neo-liberals who stress the importance of individual rights and the limitation of the state. The main divisions are indicated below.

1 *Traditional conservatives* who emphasise traditional 'Tory' interests such as hierarchy and order in opposition to socialism.
2 *Neo-liberals* (Friedman and Hayek for example) who have revived and developed classical liberal ideas such as the importance of the individual, the limited role of the state and the value of the free market. These classical liberals can be divided into the three major neo-liberal schools: Chicago, Austrian and public choice. There is a strongly libertarian vein in neo-liberalism which, at its extreme, goes over into the *minimal statism* of Nozick (1974) and the *anarcho-capitalism* of Rothbard (1978). Rather confusingly, 'neo-liberalism' has also been used in the USA to describe a group of market-orientated 'left-liberals' who have been critical of the intervention of the state (Ashford, 1991, p.186).
3 *Neo-conservatives* who, in the USA, are a group of former left liberal intellectuals who became more conservative in the 1960s. In the UK, neo-conservatism has also been used to refer to the traditional conservative revival as in 1 above.
4 *The Moral Majority* is a term used to describe groups which are linked to, but distinct from, the new right. They include religious and other fundamentalist organisations committed to moral values in public and private life. They are varied; some support an active state while others are 'libertarian' (Davies, 1991, p.188).

The Market

The neo-liberal new right generally based its ideology on ideas about individualism and the necessity for free market solutions to economic problems. There was a rejection of the social efficiency criterion that had been used to justify government intervention (Cheung, 1978; Levačić, 1987). The market was seen as an integrating force affecting society as a whole and having the potential to produce social order, justice, economic growth and higher standards of living even for the very poorest members of society. Within the economic framework, there were forces which interfered with the market's effective working, such as the imperfections imposed by organised political interests (Levitas, 1986b, p.2). Invariably, organised government was targeted as a primary source of market inefficiency.

Liberation stemmed from the reduction of the role of government (Gray, 1989). Reagan and Thatcher both favoured counter-interventionist policies (Savage and Robins, 1990) and many influential new right interest groups and research organisations exerted pressure on them to influence their ideas on important issues such as the expansion of market opportunities and the promotion of the 'property-owning democracy'.

In Britain, the Institute of Economic Affairs, the National Association of Freedom, the Centre for Policy Studies and the Adam Smith Institute were important in generating new ideas about the reform of education, social welfare and health services and the improvement of local and national government administration. How far the new right 'think tanks' influenced the Conservative party is difficult to assess precisely, but new right thinkers were widely respected as legitimate sources of inspiration on policy within the party (Levitas, 1986b). In the USA, the Heritage Foundation and the American Enterprise Institute were prominent in pursuing market-orientated policies together with a variety of other powerful new right-influenced Washington lobby groups (Meyer, 1982).

The Realities of Power

The link between theory and practice was subtle. Neither Reagan–Bush nor Thatcher–Major were ever pure ideologues (see Introduction, pp.12–13). The assimilation of new right ideas into policy was always pragmatic and related to the practicalities of political life. David Stockman, one of Ronald Reagan's senior economic advisors who headed the Office of Management and Budget up to 1985, graphically describes Reagan's response to the ideas of the supply-side economists:

> He [Reagan] leaned to the right, there was no doubt about that. Yet his conservative vision was only a vision. He had a sense of ultimate values and a feel for long-term directions, but he had no blueprint for radical governance. He had no concrete program to dislocate and traumatise the here-and-now of American society.
>
> (Stockman, 1986, p.10)

Political compromises, and the strength of the bureaucracy, helped to strengthen the tendency on the right to distrust the political system. Government departments were seen as bureaucratic monsters which could blunt the aspirations of even the best free marketeer (Hoover and Plant,

1989). While the American left looked to the legislative branch of government to act as a check on the excesses and failures of the market or to local activism (Walker, 1990), the new right saw government as a hinderance to the development of the market and the legislative branch as susceptible to the pressures of vested interests.

PUBLIC CHOICE

There was, therefore, a strong feeling on the right that politics and economics were linked; a contention which was at the centre of Marxist thought. With public choice theory it was possible to interpret political actions with reference to the actions of individuals and the aggregation of preferences. Economics and politics were both concerned with scarcity and the allocation of resources (McLean, 1987, p.9). The analysis of the rational actions of individuals could therefore provide a powerful tool as a means of understanding government resource allocations. In this, public choice theory provided much that was of value in developing a view of the role of government and the motivations and interests of public bureaucrats. Public choice theory thus moved from a 'first principles' analysis using techniques like game theory and mathematical economics to a coherent 'institutional' analysis dealing with the processes and institutional behaviour of liberal democracy (Dunleavy, 1991).

The variability and flexibility of public choice theory, combined with the power and value of many of its rational choice assumptions, contributed to its expansion. By the mid-1980s public choice was regarded as the major trend in academic political science in the USA (McKay, 1984) and a major point of reference for political science and urban research in the UK. Lovrich and Neimen argue that

> it was inevitable that interest in public choice theory would spread beyond the narrow confines of academic audiences. There is also a 'practitioner' audience far in excess of what normally exists for other forms of innovative, theoretical formulations. The current sad state of the national–international economy, the advent of a conservative administration in this country [the USA], the expansion of popular movements in several states to limit taxes, the passage of 'sunset' legislation are all, perhaps, manifestations of a loss of faith in conventional ideas about how the economy can be managed, how social problems can best be tackled (or ignored) by government, and how bureaucracies can be improved to better serve the people. Currently successful politicians, and those high-level administrators appointed by them, promise 'new, innovative approaches'. These new perspectives feature a 'lean, effective and non-threatening' governmental presence entailing less regulation, lower taxes, more localized admin-

istration of programs, and a greater reliance upon both the private sector and volunteerism among the citizens.

(Lovrich and Neiman, 1984, p.xxii)

Public Choice and the Right

The left lost the initiative in developing a rational–institutional perspective which could address the challenges posed by such issues. Therefore, while public choice methodology is 'not intrinsically tied to right-wing political values (Dunleavy, 1991, p.5), institutional public choice theory became largely associated with the right as it built upon the assumptions of liberal economics and became associated with innovative approaches to difficult problems concerning the growth and inefficiencies of government. There was also a tendency mistakenly to regard public choice as being synonymous with the wider field of rational choice. Rational or social choice theorists were in fact often critical of many of the economic assumptions of public choice (North, 1986).

Even so, in public choice theory, the right possessed a very effective rational economic critique of socialism and public intervention (Downs, 1957). The right also challenged the assumptions made by Keynesian economists about how full employment and economic expansion could be bought through government intervention (Buchanan, Wagner and Burton, 1978). Public choice broadened the field of attack. Government created inflation, maldistributed resources and produced welfare commitments which were ill-suited to the achievement of self-sustaining economic growth. In short, government was responsible for recession and low growth.

Politics was seen to be in the same parlous state as the economy; influenced by welfare-liberals and interventionist socialists. Political leaders of this kind were said to be incapable of responding to the needs of the people and guilty of pursuing policies which hampered the efficient distribution of resources to communities and led to a deleterious centralisation of governmental power and authority (Bolick, 1988). At local level, vested interests often monopolised city and town governments and followed policies which had brought localities either into financial debt or to the edge of social disorder.

Public Goods

Economics was about the scarcity of resources and their distribution. Public choice theories related political action to economic scarcity by concentrating upon the most effective mechanisms for producing wealth in capitalist society. It was argued that in allocating resources a range of 'public goods' could be identified and defined. These were goods that

23

were commonly produced by governments. They required indivisibility of production and consumption, non-rivalness and non-excludability. In other words, they were produced and consumed by society where goods of one sort would not be identified as competing against each other (for example, one unit of electricity against another) and where no person could be excluded from enjoying the good (whether they paid for it or not). Non-rivalness meant that the good was of value to an individual regardless of whether he–she was the only person enjoying it (McLean, 1987).

'Pure' public goods were of such a character that it was difficult to make people pay for them. As McLean states, if one cannot be excluded from the benefits of clean air, one is tempted to 'free-ride' and get the benefit without paying. The more the good was a pure public good, the more likely it was that free-riding would take place. This phenomenon was often said to be evident in the provision of many government-supplied public goods such as education, health services and welfare in urban communities (even where there are no 'pure' public goods).

Public choice theory was concerned about the operation of markets and their workings in situations where governments produced public goods. A nation could decide to have more central planning and a greater non-market sector, but even in relatively centrally organised situations, private markets tended to exist alongside the state sector. Conversely, even in the most privately market-orientated economies, there were substantial state sectors which regulated economic, social and even political activities (McLean, 1987, p.19). Public choice recognised that the state could offset the problems caused when private markets failed or when they were shown to be incapable of distributing resources effectively, but that said little about the underlying structural problems of capitalist markets.

Problems with Market Failure and Crisis

Despite the recognition of market failure, public choice theory lacked a theory of capitalist crisis. It was concerned with the more limited task of viewing the dangers inherent in relying upon governments to provide public goods. Public choice failed to relate to the ways in which government in modern capitalist states was important in affecting profitability, promoting growth and modifying the impacts of external market fluctuations. In this connection, the theory itself was tested by market failure (Chisholm and Kivell, 1987; Buchanan et al., 1989). In practice, a strong theoretical commitment to the free market excluded the adoption of certain measures which allowed for concerted governmental intervention during times of crisis (to boost employment for example) or to maintain international competitiveness. Politicians frequently filled the gaps left by 'pure' free market prescriptions. It was often politically expedient to fund public programmes where the political repercussions of not doing so would lead to serious economic disruption or even social disorder.

In the USA, bank failures led to federal government intervention. The savings and loans institutions faced a crisis which led the federal authorities to produce a massive bail-out package. The compromise budget agreement, produced following Bush's conflict with Congress in 1990, made little impact on the increase in federal spending. The problems in the service industries in Britain and the USA cast doubts on the longer term viability of economic growth based upon the expectation of expansion in non-industrial sectors of the economy. Therefore, at the start of the 1990s, there was a problem for public choice theory in that it did not provide satisfactory answers to the critical problems posed by recession and the 'slumps' experienced in sectors such as construction and property.

Problems with Democracy and Bureaucracy

There were also problems for public choice in presenting a viable and politically attractive approach to questions concerning local participation and 'empowerment'. Public choice theory was concerned to attribute a degree of responsibility to interest groups in increasing the negative effects of such reliance. While government provisions were open to free-riding, they were also regarded as hindering the ability of society effectively to produce welfare gains for its citizens. The groups which benefited from government thus had an interest in seeking to maximise their own resources by lobbying politicians and officials. While such interventions could benefit some, they invariably worked to the detriment of society as a whole. Lovrich and Neiman, in a strong critique, argue that many public choice theorists were therefore suspicious of both government and politics:

> On the one hand, there is the traditional high regard for individual liberty and fear of collective-choice mechanisms. There is also a technical objection to 'politics', which is seen as characteristically wasteful, ineffective and subject to corruption. Bargaining, negotiation, compromise, and understandings among interest groups all tend to be seen as irrational and destructive bases upon which to construct collective decisions . . . This scepticism, sometimes loathing, for formal government extends even to voting, which is seen as seldom producing rational, utility-maximising decisions. Government is always suspect as potentially and uniquely destructive of individual liberty or as inherently incompetent and wasteful. Consequently, public choice theorists formulate non-political concepts like externalities and public goods in order to provide 'objective', extra-

25

judgmental rules for deciding when government should intervene in the private sector.

(Lovrich and Neiman, 1984, pp.6–7)

Democratic political systems allow for preferences to be expressed through elections and political mobilisation in parties and interest groups. The suspicion of group mobilisation referred to by Lovrich and Neiman may have presented a problem for public choice theorists even though there were those who related group activity with the interests of democracy (Ostrom, 1976).

Democratic systems are 'plural' in that they allow for the representation of diverse interests. Governments provide private goods as well as public goods. These can be 'commodities' such as municipal housing or 'transfers'. Transfers are payments made by governments which are not made in return for a good or service (such as old age pensions, food stamps and so on). As with public goods, private goods represent scarce resources over which interest groups frequently come into conflict. As McLean (1987) shows, there are different ways in which groups can battle over resources even though this presents problems of aggregating and assessing the multiplicity of preferences which exist in society.

The Quest for Community

Public choice theory expressed a desire to get away from centralised and bureaucratic ways of providing social and community services. This was especially relevant in the distressed areas of the cities which appeared to conservatives to be in serious danger of social collapse, particularly with respect to their economic condition, drugs problems and Aids. For many American conservatives it was the cities which represented the worst examples of the consequences of socialism and welfare-liberalism (Bolick, 1988).

Urban decay and the actions of governments were regarded by many conservatives as being intimately linked. Public choice developed alternatives to the welfare-liberal remedies relating to urban decline, the prevalence of social inequalities and communal disorder (Bolick, 1988). The tide turned against those who advocated the rationalisation of America's complex local government system through the consolidation of small (and relatively costly) local jurisdictions. In this, conservatives were supported by local residents who frequently opposed jurisdictional reforms in favour of the prevailing fragmentation and diversity of local governments. The conservatives often defended the status quo by appealing to suburban public opinion which, especially after the troubles experienced in the cities in the 1960s, was disposed to resist any encroachment into the communities by consolidator politicians.

Ostrom and Whitaker (1974) wrote about the advantages of 'community control' over the consolidation and professionalisation of police services. For them, this was an effective way of improving the quality of police services (in this case for blacks) by enabling the community to monitor police activities through the implementation of regular opinion surveys. This reflected a view in public choice theory that saw 'community' control as a way of expressing local demands by operating outside traditional welfare-orientated interest groups.

Practical Problems and Policy

Public choice therefore accepted that the political system involved bargaining over resources and interaction between governments and organised groups. However, since the efforts of groups could lead to welfare losses for the wider society there had to be an explanation of this which related to the impact of political activity in economic terms. It was here that classical principles came into play.

The Pareto principle states that if at least one person felt better off and nobody felt worse off as the result of a change, then society as a whole had benefited. This could best be approached by the operation of the competitive market. Public choice theory claimed that the market position of interest groups enabled them to enhance their bargaining power and influence an expansion of public resources in their favour. Bureaucrats in local governments tended to expand their budgets to finance projects which benefited minority interests and this served to consolidate and expand local bureaucracies. This militated against free choice by the individual consumer and thereby called into question the claims of welfare-liberals and socialists that they were acting in defence of, for example, poor communities and distressed city districts.

However, as with the public choice view of democracy, there was room for compromise on the issue of attaining the optimal Pareto position. The possibility of ever reaching such a policy objective in reality was evidently a very distant one.

> That Pareto optimality is a relative concept is amply revealed by the sheer complexity of real world markets, where the most that can be hoped for is a movement towards the optimum. The greatest challenge for society, according to public choice analysis, is constructing those institutional and regulatory arrangements which minimise market imperfections.
>
> (McKay, 1984, pp.5–6)

In practice, this had implications for the implementation of public choice concepts in public policy. It meant that adherents of public choice could claim that they were pursuing a worthy objective but that there would be

27

disagreements about the means to the desired end. Public choice, in practice, was thus open to differing interpretations when it came to policy-making, despite the basic formal agreement which advocates may have had on objectives.

This flexibility was even used to justify certain public interventions which could be regarded as conducive to meeting market-orientated objectives. Therefore, while public choice ideas were influential, most particularly in US state and local governments for instance (Lovrich and Neiman, 1984), they were not seen as necessarily prescriptive of rigid doctrinal policy alternatives. This rendered the purist public choice approach even less relevant as a way of explaining changes in the fortunes of markets and economic levels of activity and in producing practical policy objectives to deal with changing market conditions (Hoover and Plant, 1989). The fragmentation prevalent in public choice, even in its early days, reduced its value as a way of explaining change and crisis in modern capitalist economies.

REASSESSMENT ON THE LEFT

The mis-match between theory and practice was a problem for public choice theory. The left's problems, however, were more serious in the sense that the left was theoretically challenged by the right, politically on the defensive and subjectively disorientated. The left had to come to terms with the new right and developments in post-war capitalist economies (post-industrial experiences, the change in class politics and the popularity of the market for example) which fundamentally challenged the foundations of past beliefs and doctrines (Gould, 1989). The tendency on the *social democratic and liberal left* was to adapt to the predominant market trend rather than to resort to the extension and development of a distinctively socialist–rationalist analysis.

At the end of the 1970s, *neo-Marxist urban theory* was concerned with public goods provision and questions about the most democratic way of making resource decisions. However, unlike public choice theory, neo-Marxist theories tended to view government in a broader frame by regarding public goods as important in the reproduction of capital. This was done at the expense of the development of a rational institutional perspective concerning the role of bureaucracy and the potential areas for state regulation of markets.

Neo-Marxists following from the French regulationist school, did, nevertheless, make an important contribution which provided the basis for looking at the capitalist economy in terms of its economic dynamics and the process of change which was effecting a fundamental restructuring of capital. Some of the contributions of the regulationists are useful. The use of a rationalist framework for analysing bureaucracy, combined with such a broader perspective on the changing structure of capital therefore poses a challenge in developing a multi-theoretical approach.

Collective Consumption

Economists in the 1950s had been aware of the importance of the distinction between public and private goods and laid the foundations for further theorising about that distinction (Pinch, 1985, pp.6–7). For the left, public provisions were traditionally associated with social objectives concerning either greater equity or the desire to control basic industries in the public or national interest. There was less concern with their technical characteristics as public or private goods.

The theory of 'collective consumption' was concerned with goods and services which were provided through the government sector on a non-market basis (Pinch, 1985, p.14). It provided a way of analysing the general process of public goods provision. Collective consumption included services allocated by public agencies which were supported by taxation and it covered public provisions of a partly commercial kind, such as housing, which could be subsidised by government and funded by way of the rents collected from tenants. Such goods were 'consumed' by members of society on a collective basis. This excluded the non-consumption activities of the public sector such as monetary payments in the form of welfare (or 'transfer') payments and direct actions to regulate production. Collective consumption therefore centred upon services which were collectively organised and managed by the state and which could be provided on a non-market basis or be tax-supported. This excluded commercial services which could be provided by governments. Consequently, collective consumption theories tended to be concerned with citizen access to public sector services and conflicts between groups which competed for scarce resources in housing, education, health, public transport and so on.

Collective consumption was therefore concerned with activities which were implicitly subordinated to the more profitable market sectors of capitalist economies. The interactions between the state and urban communities in this context were analysed by the French writer Manuel Castells.

The Urban System

Castells (1977, 1978) modified his view of the collective consumption process during the 1970s as he shifted his political ground towards the centre. This made it difficult to settle upon a precise definition of collective consumption, but the shift did indicate the degree to which the neo-Marxist Castells prepared the theoretical ground for later revisions to orthodox Marxism. Neo-Marxists began to find areas of agreement with social democrats who had long advocated basic reforms of the capitalist market system instead of the fundamental rejection of the 'anarchy' of the prevailing economic system.

Castells rejected the classic Marxist view that it was the economic base of society which was the ultimate and most important influence upon all the various forms of social and political relationships in society. In its place Castells substituted a three-level relationship involving the economic, political and ideological spheres. The 'system as a whole' would always involve the domination of one of the three levels depending upon the prevailing nature of society. In a competitive capitalist society, the dominant level was economic, but it was not regarded as the motive force described by Marx. Indeed, each of the three levels was relatively autonomous despite the influence of the economic level 'in the last instance' in regulating the interrelations between levels (Saunders, 1979, p.183).

Castells regarded the urban system as being an integral part of the total system involving the interactions of the economic, political and ideological spheres. In this, the urban system could be viewed as a distinctive element in society which not only reflected the three-level structure in the broader society, but also facilitated the process of collective consumption. The urban system was thus regarded as one which performed an economic function in the broader social context.

There was also a spatial aspect to the analysis which has since been criticised (Saunders, 1979). Castells argued that the urban system did not perform an economically productive role as this was performed at the regional level; the argument being that factory production was regionalised because of the distributed nature of the production process between different cities. Therefore, the urban system was said to have a function in the process of collective consumption which assisted society in sustaining its workforce. The urban system provided housing, hospitals, health services and education so as to continue the reproduction of the labour force under capitalism.

While capitalism needed to reproduce and sustain labour in this way, it was nevertheless a costly business which was not always profitable. As with the public choice theory view of public goods (but from a different political standpoint), this provided a view of the governmental provision as essentially unattractive to private capital and thus prone to government subsidisation. This was an implicit criticism in the collective consumption theory of non-market provisions which was reminiscent of that provided by conservatives who also identified the problems of subsidised community and social services.

The neo-Marxists searched for new forms of service delivery outside the public sector while conservatives prescribed privatisation as a solution to overcoming the shortcomings of public services. The disenchantment of the left with the mixed economy therefore led to a re-evaluation of why it was that private market provisions were often very efficient in economic terms and why the market could be highly responsive to consumer needs.

A New Politics?

The logic of the analysis of collective consumption and markets was consistent with the reassessment of the political strategy to be adopted by the working class. Political conflicts were related to social cleavages in society. For example, there would be competition between private and public transport users and private and public health consumers over resources in the post-industrial consumerist society. Dunleavy (1979, 1980a) suggested that such divisions would not only be expressed in classical class terms, but would arise from the differing material interests perceived by those operating in different markets. Coalitions of groups would form to campaign on certain issues related to the distribution of resources and the allocation of government subsidies. These issues would cut across traditional social class distinctions.

Similarly, Cawson and Saunders (1983) identified a sphere of consumption sector politics related to public welfare provision. The location of individuals and groups in the sphere of consumption provided a material basis for the development of political perceptions and socially related demands. This produces a variety of 'consumer groups' which may conflict with each other at the local level and defend their interests within the consumption sphere. It was a conception of local politics which implied that the social class orientation of groups had become less important than their narrower self-interests in a competitive environment and that the prospects for class unity in a plural society were extremely limited.

Cawson and Saunders, and others, stressed the need to mobilise diverse local groups by way of stimulating local democracy and control. However, this goal could be remote if Leadbeater's plea for 'progressive individualism' was taken seriously. It was left unclear as to how this could contribute towards the necessary cohesion between groups seeking unity even on the basis of relatively modest consumer-orientated demands (Leadbeater, 1988, p.14). Phal (1989) goes further by dismissing class in favour of consumption as a critical social force in development (Mullins, 1991).

Problems Concerning Pluralism and Democracy

An important consequence of the left's reorientation was the blurring of theoretical demarcations. Recognition of this was evident in the writings of pluralist theorists in the mid-1970s when it was contended that a plural and democratic political system could be envisaged both in capitalist and socialist states (as then exemplified by the eastern European communist countries). In essence, this was an argument which rested on the proposition that the 'economic base' of society was less important than the political infrastructure as a support for democracy and the freer access of groups to government (Manley, 1983).

Manley connected the political and economic aspects of pluralist societies. For Manley, the concept of pluralism had been severely tested in view of the imperfections prevalent in modern industrialised societies with democratic systems. Pluralism at one time was a reasonably coherent theory asserting that structure of political power (particularly in America) was made up of many competing group interests enjoying the fruits of economic expansion and the extension of civil and political rights (Manley, 1983, pp.368–9).

Early pluralist theory stressed the openness of the American political system and the access which interest groups enjoyed in their attempts to influence national and local governments. However, during the late 1970s and early 1980s, many pluralists were worried about a socially unequal society. America's political system appeared to provide no guarantee that economically and socially disadvantaged groups could gain access to government or significantly influence decision-makers at the national and local levels. It was more than simply 'market failure' which was responsible for the inequalities in society. That the quality of democracy in pluralist politics was affected by 'flaws and deficiencies' (Manley, 1983, p.369) was therefore recognised in pluralist literature. For Manley, this signified a re-evaluation of pluralism and the adoption of a more radical perspective in response to society's maldistribution of wealth and income and the shortcomings of liberal democracy.

Dahl and Lindblom (1976) advocated structural reforms to alleviate economic adversity and to 'democratize the American polyarchy' (Manley, 1983, p.371). They criticised the 'incapacities' and even 'perversities' of American polyarchy which were present even after years of 'opportunity' and reform. The politico-economic system

> remains both sluggish and feckless in advancing on problems on which it has the advantage of decades of experience in policy-making: poverty and maldistribution of wealth, racial unequality, health care, public education, inflation and unemployment, and industrial relations.
>
> (Dahl and Lindblom, 1976, p.21, quoted in Manley, 1983, p.372)

This focused on the structure of the political system and processes of economic wealth distribution. The emphasis upon 'structural' changes was crucial as a prescriptive element in the new version of pluralism. However, it was on this basis that pluralism could envisage the realisation of democracy in diverse societies with differing economic systems (Manley, 1983, p.378).

REGULATION THEORY AND 'NEW TIMES'

This did not provide a theory of crisis. Indeed, it blurred the issue of crisis under capitalism by introducing the possibility of a non-specific form of economic base acting as host to a vaguely defined 'responsive' plural political structure. Regulation theory and its 'post-Fordist' derivative pointed to a more clearly defined link between capitalist crisis and the political consequences of technological change which were largely conducive to sustained capitalist development.

Advanced industrial societies, such as the UK and the USA, were said to be entering a new era in which new industrial processes were replacing those used in the old manufacturing industries characteristic of the era of Henry Ford-style mass production (which had introduced the manufacture of standardised items on a mass 'production-line' basis as in the Ford car factory in the USA).

Regulation theory developed in response to the crisis of post-Second World War Keynesian economics during the 1970s (Clarke, 1988). Keynesian economics assumed that economic crises could be managed by way of fiscal policy and control of the aggregate level of demand in the economy. The regulationist view challenged the Keynesian analysis by developing a critique of the capitalist market which pointed to the inability of capitalism to ensure sustained accumulation.

Flexible Specialisation

Aglietta (1979), writing about the USA, referred to a phase of capitalism which was characterised by a new regime of accumulation. This involved the extensive development of Fordist mass production and changes in ways of integrating or handling capital accumulation. The French regulationists concentrated upon the stages in the development of capitalism from a competitive regulation through Fordist regulation to a transitional period in the 1970s onwards where flexible specialisation in industry becomes common.

Flexible specialisation signified a development away from the standardisation of mass-produced consumer goods. Technological advances and changes in industrial organisation facilitated the new approach. Under a system of flexible specialisation, skilled workers and machinery were capable of responding to the specialised needs of customers. Mass production was retained, but companies could more effectively cater for niche markets and rapidly adapt their distribution to the demands of an ever-changing market.

Capitalist Crisis

Capitalism depended upon arrangements which facilitated the process of accumulation. As capital came under pressure in the post-Second World War period, profitability was affected as prevailing modes of regulation failed. The crisis of capitalism was seen as structural, involving a critical period for the regime of accumulation characteristic of Fordism. Profitability and productivity were affected as large scale industries of the Fordist era collapsed. From this, there came the possibility of the emergence of a post-Fordist period in which capitalism would adopt new ways of trying to overcome declining profitability and periodic crisis. The development of the service sector and high technology industries facilitated dramatic improvements in productivity and profitability. New patterns of corporate organisation enabled companies to capitalise on technological innovations in communications, thus enabling companies to relocate their plants (to 'greenfield' sites for example) and more effectively to decentralise and sub-contract. They were also able to adopt more flexible policies with regard to employment, using more short-term workers and holding down non-skilled wage levels.

New Times

For the British journal *Marxism Today*, post-Fordism was a sign of 'new times' in which industrial and cultural changes were challenging the basis of Fordism in the capitalist nations of America and western Europe. Radical changes were taking place in patterns of production and consumption which were not only facilitating the expansion of capitalism, but also delivering benefits to the working class which was being segmented and ideologically disarmed by the lure of the consumer society. Capitalism had attempted to boost its profitability during the 1980s by 'reclaiming the public sector for public accumulation' (Murray, 1988, p.9) and by weakening the power of organised labour. The drive for profitability had been hastened by the introduction of new technologies and new methods of production, such as flexible specialisation, which 'stood much of the old culture on its head' (Murray, 1988, p.9).

Abandoning Centralism

As far as developing a socialist response was concerned, post-Fordism implied a fundamental reassessment which led away from discredited centralist approaches towards solutions which would provide innovative alternative strategies. Murray (1988) proposed a strategy that envisaged a 'honeycomb of decentralised, yet synthetic institutions, integrated by a common strategy, and intervening in the economy at the level of pro-

34

duction rather than in trying to vainly plan all from on high' (Murray, 1988, p.13). The proposed economic model was explicitly to be the Italian consortia and the German industrial regions as opposed to a strong central state model which had failed so conspicuously in eastern Europe and the Soviet Union. The proposed political model was to be one of democracy, based if necessary upon compromise, with a variety of groups including non-socialists.

In many ways, this 'revisionism' represented the most dramatic expression of the abandonment of classic state socialism in favour of an analysis of capitalism which was appealing even to some of the most ardent adversaries of orthodox Marxism. Indeed, the new currents on the British and American left echoed the views of those who acknowledged the benefits of the market as an efficient mechanism for distributing and allocating resources in society. In the case of *Marxism Today*, the very validity of Marxism was questioned.

'New times' politics recognised not only the limitations, but also the potential of markets. It emphasised the positive outcomes of market behaviour which were particularly evident during times of economic upswing such as occurred in the mid- and late 1980s in the USA and Britain. It also emphasised the benefits of market activity for 'individuals' and the implications for the breaking down of traditional working-class politics (Leadbeater, 1988).

CONCLUSIONS

Regulation theory broadened the debate about politics by emphasising the economic environment within which issues could be understood. It indicated that periodic crises were caused by factors other than the mismanagement of market economies. They were endemic to capitalism, and structurally inherent in the process of economic change. While regulation theory did not produce a finished theory of crisis, it drew attention to the importance of economic change and the restructuring of capital.

'New times politics' was in some ways a liberating influence (Hall and Jacques, 1989a), especially where old style Stalinist orthodoxy had acted as a suffocating barrier to the development of new ideas. Despite the useful contribution of regulation theory and 'popular' new times politics in highlighting important characteristics of the process of change under modern capitalism, new times politics tended to over-optimistically concentrate upon the ability of markets to deliver goods and services in an effective manner. This did not adequately explain the tendency of markets also to neglect certain social provisions or to severely limit access to them especially during times of downswing or acute crisis. Markets could both provide and deprive citizens of goods and services and actually severely limit 'choice'.

Public choice theory presented an effective populist challenge because it addressed important questions about the nature of public provisions in capitalist economies. Public choice theory helped to explain why public services were not efficient and why social problems afflicted the cities even after the implementation of interventionist programmes. However, public choice theory had severe limitations as a tool of analysis and as a prescriptive device. It had no theory of crisis and, in its 'pure' form, it was incapable of relating to 'real world' policy problems.

Although its rationalistic base was valuable, public choice theory began to look less impressive as a way of explaining a world beset by market failures and economic recession (see Chapters 3 and 4). It also looked less convincing as a way of explaining urban change and local economic activity (see Chapter 2). There were also significant differences of approach within the public choice school. For example, the lack of a unified choice model was exemplified by the development of both Paretian and Wicksellian approaches; both of which failed to provide the basis for theoretical unity (Wiseman, 1989).

Dunleavy's (1991, p.5) search for a redefinition of rational choice seems apposite in this context. The task for Dunleavy is to 'selectively remove or alter contestable premises in rational choice accounts and to consider the implications of replacing them with acceptable assumptions'. This task, together with the development of a satisfactory economic perspective on change and the restructuring of capital, to use Dunleavy's words relating to public choice, provides 'a most interesting intellectual challenge'. This provides the focus for discussion in Chapter 2.

2

THE ROLE OF THE STATE

This chapter considers the role of governments in relation to the economic changes that have taken place, particularly during the last two decades. Reference is made to public choice and regulation theory. An assessment is made of the changes that have resulted from new economic conditions and the role of local and national governments in influencing change. The nature of the economic constraints and pressures for change facing governments are examined.

King (1987b) argues that one of the dominant themes in the recent literature on political economy is the importance of economic processes in determining the pattern of urbanisation and the political implications of economic change in municipal life. King maintains that it is important to examine economic processes outside the immediate terrain of any particular city because external factors play an important part in shaping city political and economic structures. He points to writers who see municipal behaviour as being, 'to a significant extent', the outcome of extra-local economic, political and social processes (Harvey, 1973; Gordon, 1976; Castells, 1977; Smith, 1979; and Katznelson, 1981). For King, they all, in their different ways, stress that cities cannot be studied apart from the wider social and economic system of which they are a part.

King also alludes to those who have regarded the state as having a minimal and reactive role (Gordon, 1976) or who have adopted an essentially neo-classical market analysis (Peterson, 1981). It is thus important not to adopt a one-sided view of state action, so one is aware of the danger of underplaying the importance of market forces in affecting state responses. Global economic changes and *the combined influence of the state and capital*, King argues, have been important in affecting declining urban communities (King, 1987b, p.229).

There are regional influences relating to the growth of provincial cities, national influences which link the fate of city economies to that of the national economy, and global influences which imply that cities are affected by international economic fluctuations. Also, there are effects which stem from the ideological commitments of national political leaders, such as the adoption of privatisation as a means of stimulating economic growth and social change. Cities are thus an integral part of a nation's political and economic system, and because of this their conditions can best be understood by viewing them in the broader context. By looking at cities in this way, a more effective assessment of the implications of specific policy initiatives and the practical consequences of the development of initiatives (such as public–private partnerships for example) may be obtained.

For King, there has been a neglect of issues concerning the degree to which the state has developed specific interests which reflect its own interest in survival and certain city-wide concerns. Sub-national governments may be regarded as operating in a difficult situation. On the one hand, they take account of market forces by acting to enhance the interests of corporate investors, while at the same time they try to maintain a viable economy in declining inner-city communities (Mayer, 1991). Capital has preferred to locate away from traditionally low-skilled working-class areas with larger production plants tending to be attracted to out-of-town sites.

The Impact of Policy-Makers

The question of how far the actions of policy-makers and politicians could be responsible for national, regional and local economic growth was thus important. This can be viewed by evaluating policy initiatives. In the USA, there were criticisms from civic leaders about the Bush administration's cuts in community and regional development funding under the Community Development Block Grant (CDBG) and withdrawal of support for communities through the Economic Development Administration (EDA) which Bush proposed to abolish in the financial year (FY) 1992 federal budget (Mayer, 1991; Wollack, 1991).

However, the central and more complex questions for those looking at economic development concern the extent to which local governments are reactive to market conditions and to what degree they enjoy autonomy and freedom of action in devising policies to offset the consequences of industrial decline. King and Pierre (1990) examine these problems posed with reference to two types of local autonomy.

Type one autonomy refers to the pursuit by local government of objectives influenced by local interests and organisations, be they private, voluntary or statutory. Such interests may act as a constraint on local governments. Economic interests are at stake, and local residents (as local tax payers) may exert pressures for local politicians to keep taxes down. Public choice theory sees this as part of the market constraint upon government.

Type two autonomy refers to 'the local government's pursuit of objectives and policies independently from central government restrictions and preferences or of professional groups capable of nationalising policy options irrespective of a locality's distinctive character' (King and Pierre, 1990, p.2). Gurr and King (1987) argue that there is a propensity for local governments to derive and follow their own *distinctive interests*. In the short run, they may combine the self-interests of elected and appointed state officials. These are structured by the values and cultures of the state institutions to which they belong and by the aims of dominant political groups represented at the local level.

Gurr and King's view of local interests suggests that it is not only governments which develop specific interests when responding to economic issues. For example, economic restructuring and resource limitation have been accompanied by the prevalence of problems affecting a wide variety of non-governmental organisations. Community leaders and business leaders have raised demands to defend and promote their interests and have attempted to maximise the resources going to communities. Civic leaders have promoted their cities to create jobs and protect and maintain existing businesses (Le Gales, 1990). In practice, this has proved to be a difficult task, especially where interests conflict with those of national governments.

Partnerships have been promoted in both the UK and the USA to overcome such conflicts (see Chapter 9). The private sector has developed an important role in assisting with the development of initiatives to benefit local communities (for example, City of Hartford, 1987). Companies have been active in urban renewal and their decisions have affected the patterns of economic and social development both in distressed urban areas and central business districts (Cummings, 1988). The decisions taken by companies about the location of factories and the establishment of commercial and other ventures, are taken in conjunction with governments and public agencies. Therefore, in assessing the role of markets it is important to take account of the decisions taken by voluntary groups and private sector organisations as well as governments (see Chapters 9 and 10).

WEAK GOVERNMENT?

Public choice theory maintains that market conditions are of importance in understanding the role of governments. Governments are seen as largely weak in the face of market forces and the choices of consumers (Wiseman, 1989, p.209). In this way, public choice shares a certain 'fatalist' conception of change with post-Fordist conceptions in regulation theory.

Regulation theory provides only a partial explanation of the subtle relation between 'market forces' and the impact of government interventions. The post-Fordist pattern of economic change (or features identified with it) are said to be symptomatic of an historically significant new era of capitalist development where service industries and new techniques

39

of production and distribution have rendered traditional governmental responses (such as state intervention) both less effective and relevant as solutions to new economic problems (Murray, 1985). Governments have little influence over the location decisions of companies as flexible specialisation enables corporations to operate across national boundaries in response to changing market conditions.

Any analysis of the relationship between economic change, restructuring and the responses of governments therefore has to take account of the role of government policy strategies and the political conditions affecting change. However, it is argued here, against both the public choice and the regulationist–post-Fordist views, that governments (particularly at the local level) are not quite so inconsequential. It is accepted that they are not in a position dramatically to reverse the strongest market pressures or to avoid internationally conditioned crises and recessions. However, a balanced view of the relationship between market forces and governments is necessary to account for the strong political responses and proactive development strategies adopted at the local level (Cason, 1980; Commonwealth of Massachusetts, 1984; Mills and Young, 1985; Bartsch, 1985; Birmingham City Council, 1986) and which recognises that local governments can actually gain from market pressures which favour particular regions (King, 1990). Such 'gains' can dramatically and suddenly place depressed towns 'on the map' as far as economic development is concerned, as when Toyota decided to locate a major production plant near Derby in the British Midlands (Derbyshire County Council, 1989a, 1989b, 1989c).

The Retail Revolution: the Power of Market Forces

It is frequently argued that retail developments also produce gains for local governments which would otherwise be hard-pressed to attract other employment-generating investment. The impact of the massive 45 hectares West Edmonton Mall (WEM) in Canada provides a clear illustration of the spin-offs and attractions of the 'retail revolution'. With its 800 shops and associated leisure facilities, the WEM provided employment and acted as a catalyst for commercial and residential development in Edmonton (Blomeyer, 1988). The WEM epitomised the 'glitz' of the 1980s and the apparent ability of the market to create concept schemes that excited the public imagination in a way that no local government had ever been able to achieve.

Retail development tends to produce quick and visible development, often on an extensive scale. Out-of-town retail parks and shopping malls are a common feature of British and American urban development. The lure of retail development encouraged many British local authorities to get involved with development companies which often exerted pressures on planning committees to accept plans quickly in case investments were taken to neighbouring districts (Hannay, 1988). In the boom years of the mid- and late 1980s, market pressures drove up land prices and property returns to such an extent that local authorities were swept along on a wave of leisure and retail euphoria. The 1990s recession was a sobering reminder of the harsh effects of market activity.

PUBLIC CHOICE AND MARKET 'SLACK'

Partly because of such situations, public choice theory regards markets as highly influential and adaptive. They induce changes that are often assumed to happen with relatively little 'friction'. For example, residential choice theories influenced by Tiebout (1956) stress the importance of aggregate individual choices. As Clark and Ferguson (1983) suggest, this leads to a view of local politics which sees individual preferences being expressed by voters tending to move to local government jurisdictions that are more satisfying to them. When local polices diverge from given preferences (say, regarding the level of local taxation in an area), then the voter moves to a 'better' area. Consequently, politicians, according to this theory, should maintain distinctive policies to retain residents. Equilibrium is encouraged if cities provide all differing mixes of taxes, services and amenities which will satisfy their citizens.

Swanstrom (1988) also refers to the political dimension implicit in the attempt to assess the impact of market and economic forces. He takes a view of the political context of economic change and the development of urban policy which concentrates less specifically upon the state and more upon cities as widely varied semi-autonomous entities.

Swanstrom critically appraises Peterson's (1981) work, *City Limits*, which developed a perspective in line with public choice. Peterson argued that economic forces largely determined urban policy-making and that cities had to compete for 'mobile wealth in the market place'. Peterson argued that the permeability of urban economies placed 'limits' on redistributive policies for the poor because such policies made cities more costly for the more productive community members. Peterson therefore advocated 'developmental' policies which would provide incentives for investors to locate business activities in such cities. This investment would benefit all residents in the city since the city would enjoy the benefits of economic growth and higher employment. Economic growth was a 'unitary interest' which all citizens shared.

Swanstrom shows how this strongly market-orientated view sought to replace the pluralist view of conflict and bargaining with one that emphasised the central importance of the 'corporate image' based on consensus and technical expertise. Peterson saw developmental policies as dominating the agenda in the cities since policies were decided by the urban elite

41

responding to market forces. For Swanstrom, Peterson's theory, like pluralism, was

> based on a market model. Only in Peterson's case, instead of voters withdrawing their support at election time, the threat is that investors and residents will 'vote with their feet'. Peterson portrays the inter-governmental marketplace as taut. While allowing for some slack, Peterson restricts policy discretion to a narrow range of allocational issues which do not influence investment.
>
> (Swanstrom, 1988, p.88)

As in residential choice theories, Peterson proposed that there was a kind of free flow process at work in the market, especially where particular cities were high tax areas. This was central to the market perspective. However, according to Swanstrom, research shows that it is actually very difficult for local governments to affect investment decisions, and local tax levels have little influence over corporate location decisions. The situation is therefore not as 'taut' as the market model suggests.

While not going as far as Swanstrom in criticising the residential choice tradition, Clark and Ferguson (1983) show that the basis of individual market motivations is complicated by a number of interrelated factors. Tax burdens are important in that they can lead to migration from high tax areas, but other variables are also highly significant in influencing individual voter and corporate location decisions (Clark and Ferguson, 1983, pp.204–19). In addition, there is a net migration by non-whites into areas where there are high welfare benefits while some cities loose less population than would be expected given relatively high tax burdens.

Swanstrom argues that the impression created by Peterson and others that the economic context of cities is a prison that determines most moves that cities make

> is exaggerated. The point is not that the economic context is unim-portant but that researchers must examine the particular city and type of mobile wealth before coming to any conclusions about the effect of the economic environment on policy making. Investors do not move from one city to another with the same ease that consumers switch from one brand of toothpaste to another. There is consider-able friction in the movement of investment capital.
>
> (Swanstrom, 1988, pp. 94–5)

'Slack', rather than Peterson's market tightness, in the intergovernmental market is seen by Swanstrom to create room for internal political discretion in the cities which makes for a wide range of political influences on policy-making. Even where economic imperatives are particularly tight, development policy is 'never simply a rational response to the facts' (Swanstrom, 1988, p.95) There is thus no single agreed 'public interest':

The economic pressures on cities are real, but rarely, if ever, do they translate rationally and objectively into one best policy that represents the unitary interest of the city. Economic pressures are always mediated by political structures and political competitions.

(Swanstrom, 1988, p.107)

'REGIMES'

It is here that an analysis of different city 'regimes' may be relevant (Stone, 1976). Fainstein and Fainstein (1983a) concentrate upon the activities of the state as a whole, and that part of the state which is responsible for the setting of policy. Elected officials preside over a body of permanent officials within the more 'stable' state structure. The administration of government is susceptible to the whims of the electorate and is therefore less stable. This means that there are likely to be different political regimes responsible for the making of policy at different times.

The 'regime' approach accounts for the wide variety of differing political responses in different cities. It makes generalisations about behaviour in different cities while accounting for the important differences relating to the way in which cities respond to their particular economic environments. The type of regime that a city produces will be determined, in part, by the economic environment, but there will also be other factors such as the nature of local political alignments and the structure and workings of local power and elite groups. This may assist with the analysis of different degrees of autonomy between cities and the impacts of various styles of political leadership.

Intergovernmental Relations

In terms of a comparative analysis of Britain and the USA, the examination of the diversity of political regimes as well as policy styles (Heidenheimer et al., 1990) and governmental arrangements helps to underline the variations between local policies domestically and differences between the British and American pursuit of market-orientated policies.

There are important structural differences between the British and American systems of government (see Chapters 5 and 6) which affect the outcomes associated with market-induced changes and economic restructuring. Styles of governmental management are different in the two countries, and organisational relationships radically differ between centre and locality. Historical factors are responsible for this, combined with different philosophies about how local government should be organised and managed.

Both in the USA and in Britain, national governments took measures which modified the roles played by city governments in economic development. The Reagan years saw an attack on the ability of city governments to raise revenues and administer social programmes. Federal cutbacks over a wide range of programmes for the urban poor forced the cities to look to new sources of revenue and strike up new relations with their state governments and the private sector.

The Thatcher and Major governments were ill-disposed to 'high spending' city councils and metropolitan authorities. The attack on the ability of the big cities to raise revenues, cuts in local government expenditure and the introduction of the Community Charge and Unified Business Rate, all marked the severe limitation of local discretion over financial management. In addition, the government abolished the metropolitan level of government in London and the former metropolitan counties in favour of district-level jurisdictions. This seriously impaired the district councils in metropolitan areas in their attempts to devise conurbation-wide economic development strategies (Birmingham City Council, 1989).

The tendency for conservative administrations to curtail city governments has been strong, but there were marked differences in local attitudes and governmental responses between the UK and the USA. The Thatcher government's attack on local government represented a direct threat to the prevailing structure of intergovernmental relations by shifting decisively in favour of central government. The centre initiated policies designed to promote economic growth at the local level and in some cases took direct control of local development initiatives (through the urban development corporations). Important changes occurred in the USA, but there was a stronger resistance to the incursion of the centre, influenced by long-held notions of the constitutional rights of states and municipalities.

In economic development, the US states became more influential and this acted as a counter-balance to the pressures placed on city governments by federal cutbacks. The states operated innovative programmes which provided tangible benefits to companies. For example, Massachusetts operated its Industrial Finance Agency, a Capital Resource Company, the Community Development Finance Corporation, the Business Development Corporation, a Technology Development Corporation, a Land Bank and a Small Cities Programme (Commonwealth of Massachusetts, 1984). To some extent, the shift in favour of state governments reflected the changes taking place in the economy and the drift of industry to the suburbs. However, the attitude of federal government was in favour of this shift in power at the local level. The states were to take on some of the responsibilities formally vested with federal government, and that would lead to greater accountability over expenditure.

FISCAL STRESS AND PUBLIC CHOICE

The sphere of resource allocation raises another aspect of the question concerning public choice theory, the impact of market forces and local government autonomy. If, as public choice theorists argue, it is the market which constrains the autonomous activities of city governments, then surely the same holds with regard to the market-influenced fiscal policies adopted by cities? As in the case of economic development, it can be argued that cities are limited in what they can do by the market pressures and demands which affect them.

Clark and Ferguson (1983) challenge the assumptions of some public choice theorists by presenting an analysis of American cities which suggests that cities enjoy a wide degree of discretion in setting their fiscal policies. The cities remain partially autonomous with regard to the broader economic environment within which they work. In arguing this, Clark and Ferguson examine the view in public choice theory that there is a close relationship between industrial decline and fiscal stress in American cities. This proposes that the cities under the greatest fiscal strain are those that are tied to old and declining industries. Behind this is the public choice argument that there is a relatively straightforward linkage between the state of the market and the ability of cities to follow particular economic or fiscal policies.

Clark and Ferguson criticise Peterson's 'economic base definition' of city fiscal strain referring to a detailed statistical analysis of data on American cities. This indicates that private sector or market processes 'do not determine fiscal policies' in a simple deterministic way (Clark and Ferguson, 1983, p.76). One test of this consists of the use of inter-correlating variables commonly accepted as defining distressed or growing cities. These demonstrate that there is only a low level of correlation and that the variables measured do not 'go together' despite some limited clustering. No single dimension (such as population change, age of city, percentage of population being black, etc.) emerges to imply a link between such factors and the degree of fiscal stress in different cities. In short, cities can offset the effects of economic constraint and recession by following distinctive policies designed to take account of the specific conditions which they face.

This finding, together with an appreciation of differing political regimes, may be relevant in Britain. In the 1980s, some UK cities devised effective strategies for overcoming central constraints, while others adopted a confrontational attitude towards the Conservative government. Some local authorities coped better than others in the face of market-induced changes and regional disparities. Whereas cities such as Glasgow and Birmingham attracted companies and promoted economic initiatives, others notably failed to engender the confidence needed to inspire the necessary infusion of investment or entrepreneurial interest. Liverpool's unhappy period of conflict with the centre under its Marxist Militant Tendency leadership provided a disincentive to potential inward investors during a long period of political strife in the 1980s (Taaffe and Mulhearn, 1988). City regimes differed, even between local authorities controlled by the Labour party (see Chapter 6).

POST-FORDISM: AN IRRESISTIBLE FORCE?

Such antagonistic policies were often justified by reference to the effects of government policies and the decline of Britain's industrial base. The British left, in the early 1980s, drew a direct connection between the workings of the market and the policies of a Conservative government committed to privatisation, non-intervention and deregulation. There was a sense in which many on the left felt that the market-orientated policies of the government and the workings and consequences of the 'free market' were indicative of a process of fundamental political change and economic restructuring which were both illustrative of a new period of British capitalist development characterised by post-Fordism.

Capital and 'Deindustrialisation'

In the USA and the UK, in the 1980s and 1990s 'deindustrialisation' and unemployment provided a major focus of concern for local politicians (Moore and Richardson, 1989). Employers' organisations demanded government action to assist declining areas. Trade unions were active in defending the interests of their members when plant closures took place. Local governments sought central government assistance or, as in the UK, attempted to gain grants from the European Community to assist run-down industrial areas. 'Deindustrialisation' and unemployment were thus important points of reference when considering the interplay of governments and the market (Duggan and Mayer, 1985; Spring, 1987; Jacobs, 1990).

Regulation theorists and the post-Fordists regarded the state as important in helping to facilitate the reproduction of capital. The role of the state inevitably changed as economic restructuring took place, since the state had a role in the process of reproduction and governments exerted an influence over national economic policy (Clarke, 1988). The state thus facilitated restructuring, assisted business, supported market-led initiatives and ensured the reproduction of a skilled workforce.

This acknowledged the continued importance of the state in the process of capital accumulation. The demise of the traditional smokestack industries in places such as Detroit in the USA and the Black Country in Britain seemed to fit the post-Fordist picture well, as did visions of public agencies operating under adverse market conditions which threatened to undermine their powers or de-stabilise local economies (such as by way of a major factory closure). In such situations, local governments often had to 'stand by' as industries declined and high technology companies established themselves away from traditional city locations (Bartsch, 1985).

There were problems, particularly with 'late' regulationist (mid-1980s) and post-Fordist analyses of the 'new times' variety (see Chapter 1). Williams *et al.* (1989) present a critique of the regulationist–post-Fordist view developed by Michael Piore and Charles Sable (1984) in *The Second Industrial Divide: Possibilities for Prosperity*. The critique refers to post-industrialist tendencies in both the British and American economies, and the degree to which a post-Fordist perspective, building on regulation theory, can be applied to them.

Williams *et al.* refer to the relevance of American-based Piore and Sable's work in Britain. This is because of the importance which Piore and Sable ascribe to 'flexible specialisation'. This concern is also evident in the work of Murray (1985) and items published in *Marxism Today*. The element in Piore and Sable's work which is central to the discussion of post-Fordism in Britain and the USA relates to the conceptual distinction between mass production and flexible specialisation. They accept the characterisation of mass production involving the use of special purpose (product specific) machines and semi-skilled workers producing standardised goods. Flexible specialisation is based upon the use of skilled workers producing a variety of customised goods.

Williams *et al.* argue that Piore and Sable employ 'an ambitious superstructure' which is constructed upon this one opposition between mass production and flexible specialisation. In this, the Piore and Sable analysis refers to three interrelated themes:

1 a theory of types of economy, their characteristic problems, and how these problems can and have been resolved;
2 an interpretative history of the development of modern manufacturing since 1800 which sees the most recent period as one of a particular kind responding to the modern needs of capital;
3 an analysis of the current crisis of the advanced economies and its possible solution.

It is the analysis of these upon 'the foundation of one opposition' (between mass production and flexible specialisation) that most concerns Williams *et al.* (1989, p.164). Their argument is that Piore and Sable see mass production and flexible specialisation not only as paradigmatic types of production, but also as *historically realised types of economy* where 'one type of production dominates' over given geographic areas – regionally, nationally, or internationally (Williams *et al.*, 1989, p.164). In this way, the USA and Britain are regarded by Piore and Sable as having been mass production economies during the twentieth century, but as now moving towards a historical (post-Fordist) period of flexible specialisation, service expansion and extensive deindustrialisation.

Williams *et al.* concede that the distinction between the types of economy mentioned by Piore and Sabel are often intentionally blurred (this also applies to the wider post-Fordist literature). However, the essential two-way opposition between the types of production appears as a recurrent theme in the argument, implying that the demise of mass production is, in some ill-defined way, *historically inevitable*. This is in spite of post-Fordism's frequent denial of the notion of historically irresistible forces in history.

Diverse Economies and Patterns of Change

The conception of weak local government, set against the historically inevitable forces of post-Fordist restructuring, certainly says something about the *direction of economic change* and its political implications *under particular circumstances*. However, it provides little in the way of an understanding of the impact of specific public policy initiatives designed to arrest economic decline and promote growth in declining urban areas. Nor does it say much about the interrelation between wider economic issues and the 'real' concerns, interests and objectives of policy-makers. Indeed, there are occasions when policy-makers are concerned to reverse the decline of their city industrial bases and follow policies designed to regenerate manufacturing and industrial investment. Certain 'traditional' industries may still tend to locate or expand in areas specialising in certain skills where 'flexibility' is not necessarily a major location factor (Sheffield Development Corporation, 1989).

Empirical evidence to support the view that economies are irreversibly moving towards the application of flexible specialisation therefore remains controversial (King and Pierre, 1990, p.276). In reality a diversity of production modes may be developing giving rise to a more complex 'regime of accumulation'. In this context, policy-makers will adopt differing strategies and express different interests according to particular local economic circumstances, levels of technology and labour force characteristics. Towns with a high technology capability will tend to have different objectives and expectations from those able to capitalise upon specific local (low-tech) labour force skills as in the ceramics and textile industries for example.

Recent attempts to relate policy issues to market changes have attempted to assess the policy–market relationship. These have emphasised the relative weakness of local governments and the fragmentation of political representation. Attention has been given to how local authorities will have to develop new 'coping' strategies to deal with change (Stoker, 1990, p.260). This approach promises to provide a more subtle picture than that implied in 'classic' post-Fordism.

CONCLUSIONS

Despite the emphasis upon the influence of market changes in post-Fordist and public choice approaches, there is evidence that governments are not simply reactive to market forces even though they may be in a relatively weak position. Governments at the national and local levels play a role in effecting changes in policies that impact upon communities and economic growth and they can modify the affects of market-induced influences. Governments provide an operational environment for companies and facilitate their development strategies by providing planning permission and other support services. By taking into account the relationship between capital and the state, the role of government may be evaluated with reference to the interests pursued by capital, the state and other groups.

Governmental systems and political regimes are important. Different political dispositions create different environments for capital to relate to. Differences in organisation and style can have profound affects upon the ways in which policies work. Chapters 3 and 4 examine the implications of market and urban change, Chapters 5 and 6 cover some of the varied political responses and institutional arrangements for handling them.

3

URBAN AMERICA: HIGH TECHNOLOGY AND POVERTY

In 1991, economic recession in the UK and the USA was accompanied by a market downturn in the manufacturing and service sectors and in construction and property. The Gulf crisis, precipitated by Iraq's invasion of Kuwait in August 1990, initially led to fluctuations in oil prices which added to the pessimism about the future of the world economy and the longer term prospects for US economic growth.

In the USA, the 1991 Gulf War threatened to affect the nation's cities. There were worries about the position of Arab minorities, concern about the demands placed upon local government personnel affected by military call-up (McCarty, 1991) and anxiety over the threat of terrorism in major urban centres.

THE UK AND THE USA: UPSWING AND DOWNTURN

Inflationary pressures in the major western economies persisted despite the implementation of cautious monetary policies and the slowdown in economic growth. Capital became more internationalised despite signs that nations were moving towards protectionism. The realities of world trade were increasingly interpreted in terms of the dangers of foreign incursion into domestic markets.

The transition to recession focused attention on the fact that economic activity involved both growth and decline. Table 3.1. shows that in early 1991 the USA and the UK were facing higher unemployment, falling retail sales volume and declining industrial production. In Japan and the western part of Germany, these indicators all pointed in the opposite direction.

Interrelated Crises

An appreciation of the double edge of market activity is therefore crucial in developing an understanding of the economic and social changes taking place in British and American cities (Hall and Preston, 1988). In both countries, there were interrelated crises concerning:

1 The international challenges posed to domestic manufacturing industries (including high-tech) and cyclical fluctuations in the economy.
2 The 1990s slump in the property and construction industries.
3 Acute problems within the financial–banking sector highlighted by bank failures and the political repercussions of revelations about malpractices in the financial sector (in particular the case of the Bank of Credit and Commerce International; the fraud-ridden bank shut down by regulators in July 1991).
4 Problems associated with uneven economic development and social polarisation.

Table 3.1 UK and US comparative economic indicators: production and employment 1984–91

Country	Retail sales volume	Industrial production	Unemployment rate
United States			
1984	95.3	98.3	7.4
1985	100.0	100.0	7.1
1986	105.7	100.9	6.9
1987	108.3	105.9	6.1
1988	112.2	111.6	5.4
1989	114.7	114.5	5.2
1990	114.2	115.7	5.4
1991 (Jan.)	−6.0	−0.8	6.1
1991 (Feb.)	−4.7	−2.6	6.4
1991 (Mar.)	−	−3.3	6.8
United Kingdom			
1984	95.6	94.8	11.7
1985	100.0	100.0	11.2
1986	105.2	102.4	11.2
1987	110.7	105.7	10.3
1988	117.7	109.6	8.5
1989	119.9	110.0	7.1
1990	120.4	109.2	6.9
1991 (Jan.)	−1.5	−4.2	7.8
1991 (Feb.)	−2.6	−2.5	8.1
1991 (Mar.)	2.0	−	8.6

Source: Datastream and WEFA
Note: Data for retail sales volume and industrial production are in index form with 1985 equal to 100. Quarterly data show percentage change over corresponding period in the previous year. Unemployment is a percentage of the total labour force.

These problems were expressed in different ways in the UK and the USA (as explained in this chapter and in Chapter 4), although they were related by way of international cyclical fluctuations and market conditions (Hall and Preston, 1988). Recession also highlighted the complex interaction between the public and private sectors as governments strove to provide social supports for the worst-off (such as the unemployed) and continued to promote the interests of domestic companies. Such efforts implied public spending. For example, in the USA, the federal government directly supported the semiconductor industry. In Britain, the Major government boasted about increases in public spending on health and education.

While certain sectors of the economies of both the USA and the UK grew in the 1980s, there was a marked social polarisation between those suffering poverty and those benefiting from the development of new industrial and commercial activities (Reich, 1991). The capitalist market produced benefits which stimulated economic growth, but failed to produce substantially improved prosperity for those at the lower end of the income scale. The market was inherently prone to produce imbalances within and between sectors of the economy and frequently ceased to operate in an efficient manner. The market failed to provide resources for those who were marginalised in the process of economic change (Oppenheim, 1990).

The stock market fluctuations of the late 1980s and early 1990s were symptomatic of the dramatic up- and downswings which could occur. The 'crash' of 1987, and the 'mini crash' of 1989, sent shock waves through the financial markets of the world. The international financial markets were susceptible to speculative pressures resulting from the political developments flowing from the Gulf crisis and from the general downturn in the USA and the UK economies. These events, combined with the temperamental attitudes of financial institutions, threatened the prospects for sustained economic growth and stability.

The news that Japanese institutional investors, eager to cover their domestic stock market losses, had been heavy net sellers of US securities during early 1990, marked an about-turn of Japanese funds from the USA to Europe which cost the USA around US$20 billion during the first half of that year alone. This, at a time when there was a recession in the USA, intensified American criticisms of Japan and its strategic impact on the US economy. It also served as a reminder that market movements could render even the most buoyant economy vulnerable to forces largely beyond the control of governments.

In the UK and the USA, there was concern about the downturn in property markets contrasting with the boom years of the mid-to-late 1980s. In the USA, there was growing concern about the effects of the 1990s recession in the north east of the country. These were compounded by reports showing that the banks were facing increasing pressures following a fall in real consumer spending. The results of recession in the USA initially could be seen in property development, the air travel industry and retailing.

Constant Restructuring and Uneven Development

Uneven economic development emphasised the effects of both high growth and decline. Uneven development meant that the decline in traditional manufacturing industries was accompanied by an unprecedented level of development in high technology applications in commerce and industry, together with an expansion of the service sector. Flexible specialisation changed patterns of employment. There were changes in the spatial distribution of economic growth and resources which influenced the development of imbalances between nations, regional imbalances and the nature of change in urban areas (Hall and Preston, 1988). This involved a complex interaction of market factors which both tended to strengthen established centres of economic power (financial centres such as New York and London for example) and effected shifts of capital in the direction of newly expanding provincial or regional locations.

UNEVEN DEVELOPMENT IN THE USA

With regard to the USA, these developments had to be viewed within an international context and with reference to the structural problems of the US economy. The shift away from traditional centres of production and the relative neglect of the distressed 'central cities' (CCs) by capital was indicative of a more profound set of changes affecting the US economy. These involved the competitiveness of US industry overseas and in domestic markets and the 'invasion' of the USA by foreign capital. International corporations wanted new kinds of production facilities, new relationships with domestic producers and revised ways of employing and organising the labour force. These changes, combined with changes in the financial services sector, contributed to a radical change in the nature of the economic activities located in major urban centres, redefined the nature of interactions between these financial centres and continued and deepened the social polarisation in the cities between 'money men' and the urban poor.

'Foregone Production' and the Crisis in Manufacturing

In 'snowbelt' states such as Michigan (Bernard, 1990a), the economic slowdown began late in 1979, well before recession became marked nationally. Michigan's unemployment rate was just under 14 per cent at the end of 1981, but by December 1982 it had reached 17.3 per cent (Bartsch, 1985). The decline in manufacturing posed a threat to America's position in international markets. Competition from Japan in the automobile sector was one obvious example of this, but the penetration of foreign firms into the US market was more extensive in other fields, including electronics

53

and computers and in manufactured consumer goods. The Reagan approach to stimulating the economy through the application of market criteria was very appealing to many Americans who wanted to see a strengthening of the US position in such markets and who sought to regenerate declining domestic industries. The 'release' of resources for investment which would 'trickle-down' into the economy would therefore be the key to growth and, by implication, would lead to the elimination of economic inefficiency and waste in the 'soft' areas dependent upon public subsidisation.

Falling inflation after 1980 provided an impetus for this strategy. By 1982, consumer prices were rising at only about 4 per cent annually and interest rates began to fall. By 1983, the economy was beginning to pull out of the depths of recession. Unemployment began to fall as companies became more confident and as personal incomes began to increase. However, the administration continued to emphasise the need for tax reductions, particularly for businesses.

The reduction in inflation was, therefore, achieved at some cost. The recession helped to reduce inflationary pressures, but 'foregone output' produced substantial losses of income for workers and companies together with bankruptcies and job losses (Sawhill and Stone, 1984). Precise calculations of the magnitude of 'foregone production' were difficult to make, although one indicator was the calculation of losses of after-tax income per household. The calculation made by Sawhill and Stone for The Urban Institute showed that had there been no recession in 1981–82, the average person in the USA would have accumulated over US$1000 in additional income between 1981 and 1983. The implication of this was that, for a wide range of social classes, the recession had a longer term dampening effect on real living standards despite the reduction in inflation.

Even after substantial restructuring, some major USA manufacturing companies at the start of the 1990s were still facing problems in their attempts to remain competitive. For example, General Motors (GM) saw a significant decline in its share of the US car market which fell from 46 per cent in 1980 to 35 per cent in 1989. This was evident at a time when Japanese car makers were set to further boost their production of cars in the USA through their so-called 'transplants' – Japanese manufacturing plants within the US domestic market.

The Japanese transplants were expected to produce over two million cars in Canada and the USA by the middle of the 1990s in addition to the cars already imported into North America. Using the most advanced high technology production methods, the Japanese plants were able to undercut domestic car manufacturers in cost terms, despite domestic attempts to boost production and adopt new working practices. However, much of the Japanese 'know-how' was acquired through joint ventures with American companies. General Motors, for example, entered a joint venture with Toyota near San Francisco in California. The joint venture company (known as New United Motor Manufacturing Inc. [NUMMI]) employed 2, 800 people at the end of 1989 and could produce around 60 cars per hour.

Internationalisation in American Markets

Despite the fact that such a challenge to US industry was felt right across the range of US manufacturing, it was the internationalisation in American goods and capital markets which probably helped the country to avoid a major recession in the later 1980s. During the Reagan years, the import share of gross national product (GNP) rose from 11 per cent to 20 per cent. Foreign investment in the USA grew from 18 per cent to a massive 34 per cent of GNP.

The interpenetration of the US market by foreign companies was a major political concern. A report in *The Economist* (16 December 1989) showed that, in 1990, 'foreigners will own more of the American economy than America owns of theirs' – that there would thus be an imbalance against the USA in terms of international asset ownership. Most foreign direct investment (FDI) took the form of acquisitions and, in 1988, foreigners spent US$60 billion on acquiring American companies and only US$5 billion on starting new businesses. Indeed, the flow of FDI had been increasing substantially, particularly since the mid-1980s. Between 1977 and 1987, the foreign-controlled share of the assets employed in American manufacturing went up from 5 per cent to 12 per cent, with Britain being by far the largest investor in the USA. At the end of the 1980s, Britain held around 30 per cent of foreign-owned assets, with Japan second with 16 per cent. This was at a time when the stock of foreign investment in the USA was roughly as large as outstanding US direct investment abroad. In this sense, the USA was becoming a 'host' country for multinational enterprises and investment (see also OECD, 1987; King, 1990).

This had implications concerning the ways that companies operating internationally could affect the US economy. The 'classic' multinational company involved nationally based companies in the fostering of a portfolio of companies each responding to their own operating environments. This usually involved the 'local' companies responding to local needs. However, as Bartlett and Ghoshal (1989) argue, most major US companies started to transfer innovative products from the most developed to the less developed markets while Japanese companies internationalised through an export-based strategy manufacturing standardised products to be shipped worldwide (the 'global' company).

In the 1990s these 'models' of international company activity were becoming increasingly outdated. A new competitive situation emerged which required companies to develop the flexibility of the classic multinational to respond to local needs and be globally competitive to capture efficiencies of scale. This required a worldwide learning ability that results in wordwide innovation (Bartlett and Goshal, 1989). Together, these developments led to the emergence of 'transnational' companies where management increasingly treated worldwide operations as an integrated strategic whole. Such strategic considerations demanded that companies be ready to divert resources from one place to another as appropriate to promote the interests of the worldwide organisation. Company global competitiveness was thus becoming more attuned to the conditions of the world economy and assured through the transmission of information across global units of the transnational company.

Financial Services – The Global Market

Multinational/transnational companies were therefore keen to operate through the well-established financial markets in the USA. Corporate investments relied upon established financial centres where banks and other financial institutions could cater for the needs of internationally mobile investment flows. In the 1980s, the business districts of major American cities saw the visible signs of international corporate inward investment and the expansion of the financial services sector. Office buildings rose to accommodate foreign companies, financial institutions and legal firms benefiting from the surge in business.

High technology enabled the corporations to transact deals nationwide and keep in constant contact with production and distribution facilities located away from the city centres. Financial services were provided within the global market where companies needed to be in constant touch with market changes in London, Frankfurt, New York and Tokyo. Traditional stock exchange trading gave way to high technology dealing which sped transactions and facilitated rapid responses to share price fluctuations and market trends.

Cities such as Chicago and New York were able to capitalise on the growth of the financial services sector and the opportunities provided by the application of new technologies. New York strove to remain internationally competitive as Chicago attracted foreign banks and brokers (including the Japanese) and dramatically expanded its futures exchanges. Its three futures and options exchanges accounted for 60 per cent of worldwide business by 1989 while its equities market had grown to become a major regional and international trader.

New technologies no longer absolutely required companies to be located in traditional financial centres, so there was a noticeable locational 'fluidity' in financial services which enabled companies not only to locate in provincial centres, but also to transfer selected activities to newly emerging financial centres. Therefore, London faced increasing competition from

Paris and Frankfurt in the provision of financial services. In turn, Dublin attempted to attract business by orientating towards the London financial market. The southern cities in the USA created their own influential financial service sectors while smaller centres such as Hartford in Connect-icut built upon their specialist provisions in fields such as insurance, banking and legal services.

REGIONAL IMBALANCES: THE UNITED STATES

In the USA, a consideration of the pressures of change and their conse-quences reveals important shifts in the distribution of population and the directional flow of capital. These have affected urban growth and the complex interactions between metropolitan and rural areas.

There are important regional differences and different economic tempos in each of the major US cities. These differences were magnified as private capital sought to expand in areas of the economy which were rapidly growing, as in the semiconductor industry for example. The shift towards the expanding sectors helped to underscore the decline of traditional manu-facturing industries. The automobile industry was one that was seriously hit by the early 1980s recession. The industry's demise seriously affected motor manufacturing cities such as Detroit where unemployment and urban decay accompanied the run-down of production. Detroit had been one of the high growth rate cities of the 1950s and 1960s, but its later record of industrial failures and unemployment turned it, and the state of Michigan, into depressed areas in need of large investments in state govern-ment job creation programmes (Bartsch, 1985). This was a pattern repeated in other towns and cities, particularly those in the Mid-west and north-east of the USA.

The so-called 'Massachusetts miracle' was evidence that there could be dramatic revivals in relatively localised economies based upon the new growth industries or where there was a substantial level of government purchasing from companies such as those in the defence sector. This underscored the varied pattern of economic activity both within and between regions.

Measures of regional shares of national personal income indicated that the north-east recorded the country's largest gains in personal income between 1980 and 1986, but these were the result of growth in Vermont, New Jersey, Connecticut and Massachusetts (Bernard, 1990b, p.4) where financial services and high-tech companies were tending to locate. The overall share of national personal income from manufacturing and services in the north-east declined dramatically between 1946 and 1982 (from 42 per cent to 30 per cent) while it increased from 17 per cent to 37 per cent during the same period in the sunbelt (Bernard, 1990b, p.5).

'Counterurbanisation'?

The fact that the experience of decline was not evenly distributed regionally was also evidenced by reference to regional population shifts and the relative successes of some areas. William Frey (1989) provides an up-to-date and extremely clear perspective on US regional population changes which indicates of the pattern of economic growth and decline in cities and regions.

Frey's argument is that population redistribution in the USA has historically involved a westward regional shift as the eastern seaboard continued to 'fill in' its unsettled territory. Of the four regions (north east, midwest, south and west – see Figure 3.1), the west consistently expanded at a higher rate than the others. In addition, there was a redistribution of population to metropolitan areas and the cities. This was accompanied by a redistribution 'up the size hierarchy' as the process of urbanisation evolved (Frey, 1989, p.34). The larger metropolitan areas gained in their migration exchanges with smaller metropolitan areas and non-metropolitan areas. Immigration into the USA served to strengthen this pattern as the largest industrial centres (Boston, New York, Detroit, Chicago, etc.), located especially in the northeast and midwest, expanded.

Although this general pattern was maintained until recently, changes appeared in the 1970s. These were of such significance that Frey calls the 1970s a 'transition decade'. This was because regional redistribution patterns altered so that the south, in addition to the west, showed high rates of population growth. This was in stark contrast to past periods when the south had been considered to be 'something of a lagging region' (Frey, 1989, p.34; see also Figure 3.2).

Frey also considers the changes in urban areas during the 1970s. For the first time since industrialisation, the population of the nation's 'non-metropolitan territory' grew faster than that of its 'metropolitan territory'. This is a process which is referred to by Frey as 'counterurbanisation'. This is a misleading term in that it places too much emphasis on the outflows from traditional urban centres. Outflow, however, involves a process of urban-style expansion in the rural or suburban areas. In this way, the rural and suburban areas can be said to be urbanising; adopting urban forms in development patterns, population concentration, architectural style and aspirations.

The long-standing 'up-the-size hierarchy redistribution' that occurred across metropolitan areas also began to reverse in many places. Some of the largest industrial metropolitan areas 'sustained losses from their exchanges with smaller and non-metropolitan areas, leading to unprecedented population declines in these large areas' (Frey, 1989, p.35). This was while traditional industrial areas were suffering the effects of the decline of their industrial bases and the movements of population away

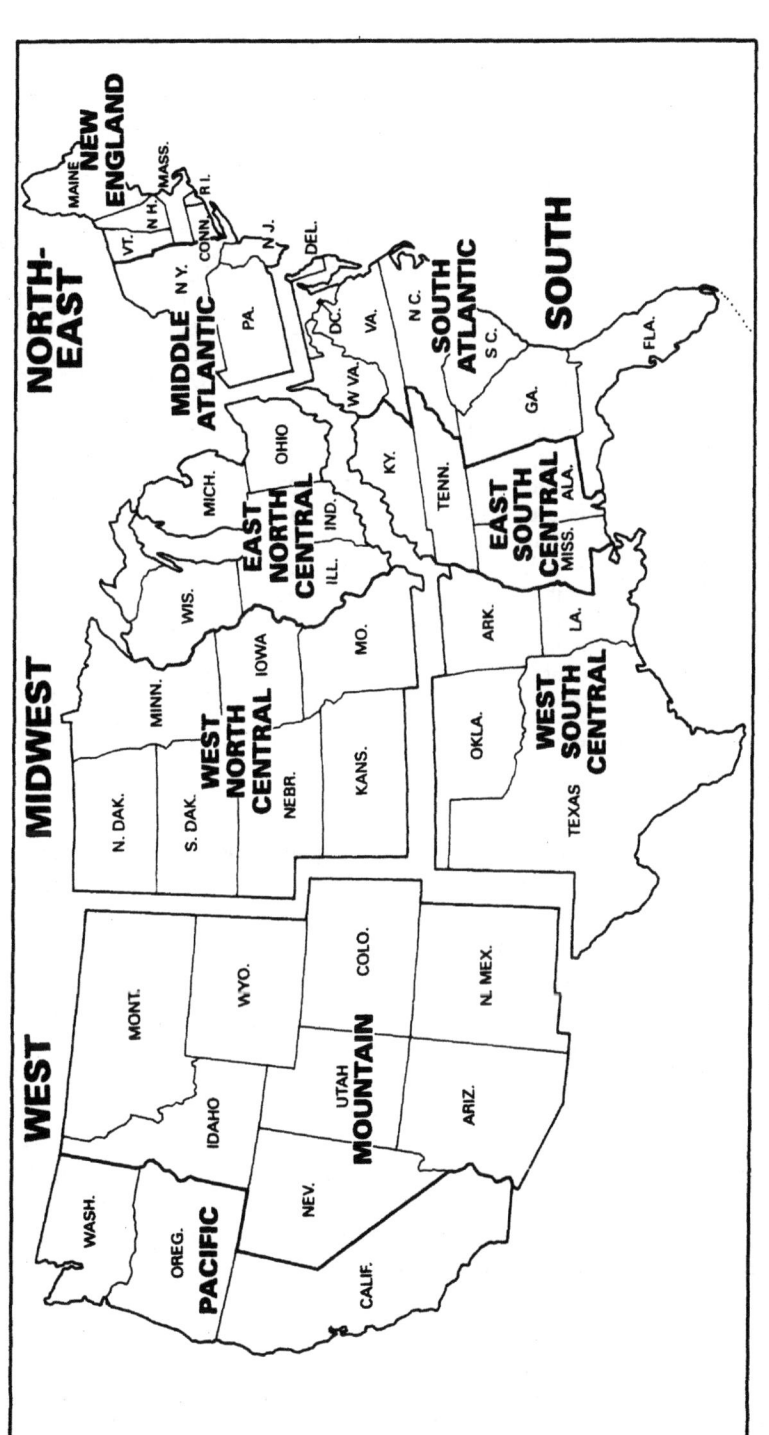

Figure 3.1 Census regions and divisions of the USA
Note: Alaska and Hawaii come under the Pacific

Table 3.2 Largest cities in the USA: populations 1980–90

Rank 1984	Rank 1990	City/State	Population 1980	1984	Estimate 1990
1	1	New York, NY	7,071,639	7,164,742	7,322,564
2	2	Los Angeles, CA	2,968,579	3,096,721	3,485,398
3	3	Chicago, IL	3,005,072	2,992,472	2,783,726
4	4	Houston, TX	1,595,138	1,705,697	1,630,553
5	5	Philadelphia, PA	1,688,210	1,646,713	1,585,577
8	6	San Diego, CA	875,538	960,452	1,110,549
6	7	Detroit, M	1,203,369	1,088,973	1,027,974
7	8	Dallas, TX	904,599	974,234	1,006,877
9	9	Phoenix, AZ	790,160	853,266	983,403
10	10	San Antonio, TX	785,927	842,779	935,933
14	11	San Jose, CA	629,402	686,178	782,248
13	12	Indianapolis, IN	700,807	710,280	741,952
11	13	Baltimore, MD	786,741	763,570	736,014
12	14	San Francisco, CA	678,974	712,753	723,959
18	15	Jacksonville, FL	540,920	577,971	672,971
20	16	Columbus, OH	565,032	566,114	632,910
17	17	Milwaukee, WI	636,298	620,811	628,088
15	18	Memphis, TN	646,170	648,399	610,337
16	19	Washington DC	638,432	622,823	606,900
19	20	Boston, MA	562,994	570,719	574,283

Source: based on figures provided by the US Bureau of the Census 1984 and 1990 (pending adjustments)

from neighbourhoods that were rapidly degenerating in terms of crime, racial tensions and infrastructural decay (Bernard, 1990). These changes resulted in altered ranking positions of the largest US cities (see Table 3.2). Growth in the wider metropolitan conurbations is shown in Table 3.3. High population growth areas such as San Jose and San Diego are notable in the individual city stakes, as are Atlanta, Dallas–Fort Worth and the continued expansion of greater Los Angeles on the metro-area list.

Judd (1983) suggests that economic stagnation during the early 1980s hit hardest the cities which were already stagnating at the fastest rates. The position of the older 'central cities' (CCs) was changing as companies moved their sights south and south-west to the so-called 'sunbelt' which, since the late 1960s, had experienced dramatic economic growth. As the Fainsteins comment, the old industrial cities

> became poorer and blacker, albeit smaller, as middle-class whites moved out to be replaced by minority households. Accompanying these well-known social changes was the relative economic decline of CCs unable to replace obsolete industries and the associated disin-

Table 3.3 The Largest Twenty Metropolitan Areas in the USA[1]

Metropolitan area		Population		Personal income per capita 1987 (dol.)	Percentage unemployed 1988
		1988 total (1,000)	Percentage change 1980–88		
New York–Northern New Jersey–Long Island, NY–NJ–CT	CMSA	18,120	3.3	20,513	3.9[2]
Los Angeles–Anaheim–Riverside, CA	CMSA	13,770	19.8	18,027	4.7
Chicago–Gary–Lake County (IL), IL–IN–WI	CMSA	8,181	3.1	17,480	6.1
San Francisco–Oakland–San Jose, CA	CMSA	6,042	12.6	21,086	4.2[3]
Philadelphia–Wilmington–Trenton, PA–NJ–DE–MD	CMSA	5,963	5.0	17,409	4.1
Detroit–Ann Arbor, MI	CMSA	4,620	–2.8	17,340	7.4
Dallas–Fort Worth, TX	CMSA	3,766	28.5	17,115	5.9
Boston–Lawrence–Salem–Lowell–Brockton, MA	NECMA	3,736	2.0	20,239	3.0
Washington, DC–MD–VA	MSA	3,734	14.9	21,539	2.9
Houston–Galveston–Brazoria, TX	CMSA	3,642	17.5	15,108	7.0
Miami–Fort Lauderdale, FL	CMSA	3,001	13.5	17,086	4.9
Cleveland–Akron–Lorain, OH	CMSA	2,769	–2.3	16,141	5.6[4]
Atlanta, GA	MSA	2,737	28.0	17,293	5.1
St Louis, MO–IL	MSA	2,467	3.8	16,706	6.0
Seattle–Tacoma, WA	CMSA	2,421	15.6	17,539	4.8[5]
Minneapolis–St Paul, MN–WI	MSA	2,388	11.7	18,277	3.4
San Diego, CA	MSA	2,370	27.3	16,633	4.3
Baltimore, MD	MSA	2,343	6.5	17,776	4.9
Pittsburgh–Beaver Valley, PA	CMSA	2,284	–5.7	15,207	6.0
Phoenix, AZ	MSA	2,030	34.5	16,077	5.1

Source: US Department of Commerce Bureau of the Census, *Statistical Abstract of the United States 1990*
Notes: [1] As defined 30 June 1989; CMSA = Consolidated metropolitan statistical area, NECMA = New England county metropolitan area, MSA = Metropolitan statistical area.
[2] Excludes Danbury and Norwalk, CT PMSA; data not available.
[3] Excludes Santa Cruz PMSA; data not available.
[4] Excludes Lorain–Elyria PMSA; data not available.
[5] Excludes Tacoma PMSA; data not available.

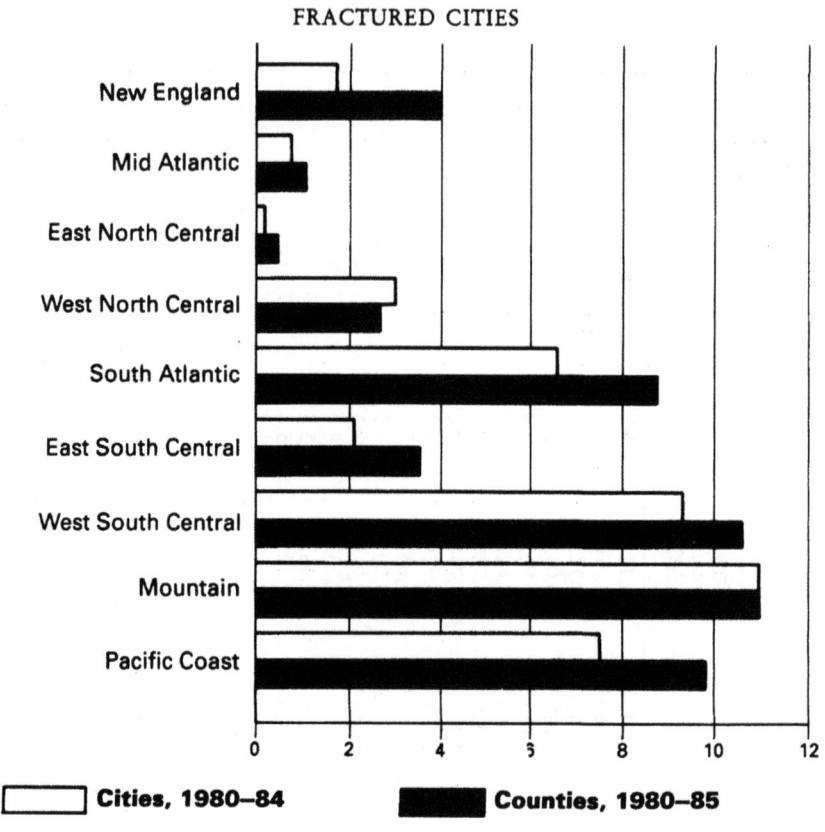

New England
Mid Atlantic
East North Central
West North Central
South Atlantic
East South Central
West South Central
Mountain
Pacific Coast

0   2   4   5   8   10   12

☐ Cities, 1980–84   ■ Counties, 1980–85

Figure 3.2 Percentage population growth in geographic divisions
Source: US Bureau of the Census

vestment from their built environments. On the whole, older American cities became vehicles for the encapsulation of minority groups and low-income whites in obsolete sectors of the economy and deteriorating physical environments.

(Fainstein and Fainstein, 1983b, p.4)

The Fainsteins showed that between 1960 and 1978, the CCs lost 9 per cent of their white populations, while their black populations grew by 40 per cent. By 1980, 55 per cent of blacks, compared with 24 per cent of whites, lived in the central cities, and nearly 33 per cent of the entire central city population was black or hispanic. Between 1969 and 1976, government-defined poverty increased by 6 per cent in CCs in general, and 16 per cent in CCs with populations larger than one million. This left 31 per cent of blacks living in CCs of any size in 1976 as officially defined as poor.

Suburban 'Urbanisation'

Cervero (1989) points to the dramatic changes which the high technology revolution brought for America's urban fabric. The variety of developments away from traditional urban centres took many forms. Cervero's classification is made with reference to the development of the following new suburban employment centres (SECs) in the USA:

1 *Office parks*, characterised by low density development and clusters of large corporate offices in parkland settings.
2 *Office centres and concentrations*, which tend to be less controlled and planned than the office parks and more agglomerative with a number of free-standing office buildings often forming 'office corridors' (such as Greenway Plaza near Houston or Greenwoood Plaza near Denver).
3 *Large-scale mixed-use developments (MXDs)*, featuring a mix of activities which encompass fairly large areas. These are usually recognised as being 'primary growth magnets' within their respective regions. Some have developed along freeways and have taken on a corridor form (for example, along the Edina-Interstate 494 south of Minneapolis and Oak Brook-Interstate 88 west of Chicago). Many resemble 'sub-cities'.
4 *Moderate-scale mixed-use developments (MXDs)*, which are similar to the large-scale variety but with smaller acreage.
5 *Sub-cities*, which are places that have been called 'urban villages', 'mega-clusters', 'suburban downtowns', and 'satellite cities'. They are noted for being downtown-like in their densities and land-use mixes, yet they retain suburban-like characteristics (zoning controls, white-collar employment, good parking facilities and so on). Most of the sub-cities are fairly new (such as the Perimeter Center north of Atlanta or the Denver Technological Center). To qualify as a sub-city, an SEC must have over ten thousand office workers and over five million square feet of office and commercial floorspace combined with fairly high average densities and a major regional shopping mall. There will also be a convention centre, on-site housing and downtown-style facilities.
6 *Large-scale office growth corridors*, which are in many ways a separate type of development to the other SECs. They represent less of an agglomeration and more of a large swath of office and mixed development within an urbanised region or state. All the large office corridors focus on one or more freeways or major highways and are inter-linked to various unrelated office parks, retail centres, housing and commercial strips.

A NEW POLITICS

Judd provides a graphic description of Denver in Colorado during the years of its most dramatic growth:

> Denver is the sunbelt city personified. Money and people for oil and gas development throughout the West are funnelled through the corporate offices which occupy the new aluminium cylinders.... Southeast of the city, all round the appropriately named Denver Tech Center, a 'silicon valley' second only to California's Bay Area electronics industry is springing up. In the last two decades, hundreds of companies moved to the Front Range, then a hundred miles along the east face of the Rocky Mountains. Many of the companies are high-tech firms specialising in military production, electronics and energy development. In 1957, Martin Marietta Aeronautics chose an area a few miles south of Denver to build the new Titan intercontinental ballistics missile. In the next few years, Beech Aircraft, IBM, Hewlett Packard, Johns-Manville, and Honeywell moved major operations to the Front Range. Professional and technical jobs proliferated, exceeded in occupational growth only by openings for clerical and service workers.... Denver's population was young and affluent. The fastest growing group was made up of people twenty-five to thirty-four years old. Even so, between 1970 and 1976, the population of children under fifteen years of age declined, showing an increasing proportion of young adults were unmarried singles or childless couples. Compared to the largest twenty-five urban areas, the Denver Standard Metropolitan Statistical Area (SMSA) ranked fifth in the growth of personal income. Probably because so many people were affluent and childless, Denver took on characteristics often associated with Southern California: a proliferation of health clubs, movie theatres, radio stations (more per capita than in any other metropolitan area in the USA), discos and 'fern bars', skylight and stained-glass restaurants, and hot tubs. In three decades, Denver shucked its cowtown image, and became 'trendy'. It became a sunbelt city, in culture and lifestyle as well as in weather.
>
> (Judd, 1983, 167-9)

Many saw this as evidence of the realisation of the American dream in which economic expansion seemed to provide an environment which sustained a lifestyle free from the problems associated with the old declined cities like Detroit. A similar ideal could also be said to apply in Britain where the cities came to compete with each other by extolling the kind of post-industrial 'virtues' of a new lifestyle mentioned in the above quote. This pattern of expansion thus typified the desire to achieve economic growth by generating a momentum that would produce professional jobs (especially in high technology and the defence industries) and 'spin-offs' in an expanding service sector, thereby creating employment and further expansion.

Reich (1991) sees similar developments in other employment areas in the USA. There has been an increase in professional jobs such as in the legal profession. Reich also shows that there has been a large increase in 'person-service jobs' in the leisure and retail sectors. These have increased while there have been dramatic reductions in employment in heavy industries like steel making and automobile manufacture.

New 'Virtues'

In social terms, the professional and executive managers of international companies and financial institutions found their niche in the better-off residential districts of the major cities close to the corporate headquarters of domestic companies and the 'branch offices' of foreign multinationals. High technology financial services were increasingly mobile with respect to their locational requirements.

The success of those employed in the tertiary sector served to highlight the plight of those living in urban areas suffering from economic decline. The flexibility of financial companies and the comparative affluence of the booming central business districts in the big metropolitan centres of America stood in stark contrast to the problems of communities facing poverty and unemployment. The problem was emphasised by the proximity of the affluent business districts to the depressed inner-city areas.

The politics generated by these changes was one that stressed the virtues of competition, the market, innovation and adaptation. It was a politics which placed emphasis on the notion of company success and the importance of specialised teams working for that success. The individual techno-expert stood to gain if he or she was part of the winning team. It was a politics far removed from collectivism, the smokestack unionism of Fordist mass production and the defensive politics of inner-urban decline and the campaigns of poorly resourced community organisations. The managerialism of the high-tech sector fostered a culture which was free of trade unions, frequently based upon incentive- and performance-related pay and attuned to rapid alterations in market conditions. The internationalisation of high-tech manufacturing complicated the lines of managerial responsibility and broke down traditional plant-based conceptions of workers' organisation. In many cases, US manufacturers had transferred production to countries where there was non-union labour and lower costs of social insurance for workers (Reich, 1991). This was a politics which was part of a culture engendered by the dynamic of the market and the quest for new technological horizons.

'Green Field' High-Tech

The high technology industries represented an attractive aspect of economic change as far as politicians and corporate decision-makers were concerned. For the corporate private sector, this was not simply because high technology seemed to represent everything that was dynamic in the economy. As Aglietta noted as early as 1976, there were other more important reasons:

> Automatic production control divides up the production process in a new way, linked on the one hand to a centralization that permits control of production by long-distance transmission of information, and on the other hand to a rearrangement of the segments of a complex labour process to allow for important savings in transfer time, quality control and the preparation of production programmes. A far more advanced centralization of production becomes compatible with a geographical decentralization of operative units (manufacture and assembly).
>
> (Aglietta, 1979, p.127)

For Aglietta, such developments benefited capitalism by allowing a greater flexibility in the installation of production units, the break-up of large concentrations of the working class and the consequent reduction of trade union organisation. In addition, capitalism could achieve economies of production and more effectively control high technology applications, programming, research and information processing. In this way, according to Aglietta, the high technology 'revolution' served to strengthen the regime of capitalist accumulation.

There was a variety of location patterns in the high technology sector. Some companies found it convenient to locate near to existing metropolitan areas where there were pre-existing high technology support structures in the economy (for example, near metropolitan Boston with its academic research institutions). However, the space requirements of computer manufacturing compelled some companies to move further away from costly urban areas (Browne, 1988, p.220).

The tendency towards high technology industry clustering was exemplified by Massachusetts and Silicon Valley in California. Each area began to 'take off' after the Second World War with heavy support coming from the military and space programmes. In the Bay Area near San Francisco, 'high tech replaced apricot and walnut orchards rather than textile mills' (Dorfman, 1988, p.263). On the east coast, at the end of the Second World War, Massachusetts contained the nation's most distinguished combination of academic-based laboratories and highly sophisticated manufacturing establishments, particularly those of Raytheon and General Electric. In California, Santa Clara County, which later became known as Silicon

Valley, was near to the growing aircraft industry of the west coast which was the only customer for integrated circuits in the years immediately after the war. It was thus during the late 1940s and early 1950s that these two locations began to lay the foundations for the high technology 'explosion' of the 1970s and 1980s.

In Massachusetts, it was Ratheon that was the major open market manufacturer in the 1950s. The wave of technological innovation in the 1950s produced important changes in the computer industry and led to the growth of new companies able to provide specialist inputs into the sector. Kenneth Olson was the company that 'set the seeds of the minicomputer boom' (Dorfman, 1988, p.267) and was the first successful manufacturer of such equipment. The location by Route 128 near Boston was a calculated decision, taking account of the proximity of the existing research infrastructure and the closeness of major related manufacturers. Similarly, the dynamic growth of high technology in Silicon Valley was directly related to the high technology expertise in the area and the influence of 'agglomeration externalities' and the 'centripetal force of spin-offs in moulding the spatial concentration of firms in the new electronics industries' (Dorfman, 1988, p.268).

CRISIS AND RECESSION

The high technology industries were closely associated with the expansive side of capitalism. The expansion of industries such as advertising, financial services and property development was also remarkable both in terms of the fortunes that could be made and the social and cultural forms to which a buoyant market gave birth. Even so, such developments during the boom years of the 1980s created the conditions for what was termed in the computer industry as the crisis of the 1990s (Cane, 1991). This was rooted in the intensity of competition between companies, their quest for profitability during a period of slackening demand and the growing importance of service suppliers using disparately sourced systems.

The Crisis in the Computing and Semiconductor Industry

The early growth of computing and semiconductor companies displayed a pattern which pointed to a close relationship between them and more traditional electronics manufacturing. Moreover, there were a series of interdependencies between such companies which remained into the 1990s and which was complicated by the internationalisation of the high technology sector (Lampe, 1988).

American computer manufacturers and software companies became acutely concerned about the lack of assurance about the future development of the sector in the face of competition, especially from Japan. In this, the high technology industries were not isolated from the variations in market conditions in the wider economy, nor were they protected against the market by their flexibility. They were therefore prone to shedding labour in adverse times as were other manufacturing and service industries.

There was an important sectoral problem. The attractiveness of high technology as against traditional 'smokestack' industrial investment could be seen in terms of the costs to capital. There was a distinction between manufacturers of high-tech goods and service providers. Labour costs accounted for a much higher proportion of revenues in the service sector, as service establishments did not have to produce large quantities of materials or maintain large physical plants (Browne, 1988, p.219). Payroll costs absorbed around 40 per cent of the revenues of management consulting and computer and data processing firms, compared to about 25 per cent of the revenues of high technology manufacturers and 20 per cent of all manufacturing revenues (Browne, 1988, p.219). Economies of scale therefore tended to be smaller for service firms than manufacturers; a factor which was reflected in the size of computer and management services establishments. The high technology-related service companies were smaller than high technology manufacturing companies (such as IBM). The relatively small size of computer and management services companies gave them a significant degree of flexibility to adapt to changing market conditions and to orientate towards new market trends. However, it made them vulnerable, especially when there was a downturn in the market or when the larger manufacturing companies came under pressure (as happened in the early 1990s). However, part of the flexibility in services of this kind was the ability to select relatively small strategically located sites, often where other companies of a similar kind had tended to cluster. Such companies, because of the nature of their business, did not have to worry about the transportation of large items of equipment or the disposal of large amounts of waste products.

CRISIS AND INTERVENTION

Worries were expressed about the future competitiveness of US high-tech companies. *The Economist* (16 December 1989) alluded to the poor productivity growth in US manufacturing when compared with the nation's international competitors and the associated problems building up in the American electronics industry. By 1989, Japan was producing more than half the world's semiconductors and nearly 80 per cent of the world's dynamic random access memory chips (DRAMs). Japan had out-produced the USA in semiconductors while the USA had fallen to below 40 per cent of world production. The most dramatic percentage decline had been in DRAMs where the USA had once accounted for the whole of the world's supply.

Many people in the electronics and computing industries called for government to 'come to the rescue'. There was already substantial support given to high technology through government military and space programmes. Congressional revelations showed that in 1989 US$40 billion was spent on research and development by the Pentagon with around US$3 billion going into high technology applications. Some Congressmen wanted to increase the research and development supported by federal government to boost the USA's share of the world semiconductor market. Free marketeers were critical of intervention for fear of 'coddling' the industry and providing the USA with a government-dependent research and development commitment in contrast to Japan's which was mainly private sector-based. The interventionists were able to point to the substantial backing of the Japanese government through its Ministry of International Trade and Industry (MITI) which had probably been one of the most significant factors in creating a supportive, expansionist and innovative environment for high technology companies.

The development of collaborative ventures between international companies, meant that the structure of the US computer and semiconductor industry changed in the 1990s in the face of the ever-present Japanese challenge (Ryan, 1990; Robinson, 1990). By the end of the 1980s, Japanese domination of the DRAM market was being challenged at a time when the internal market in Japan was becoming more competitive. Intel of California and NMB of Japan formed a partnership, with NMB benefiting from Intel's American marketing experience. Companies in South Korea (such as Samsung, Lucky, Goldstar and Hyundai) were also mounting strong challenges to the Japanese in the DRAM market and adding to world DRAM production on a substantial scale. The result was the dramatic lowering of the price of memory chips which produced less certain conditions in the market (Ryan, 1990).

Profitability had to be sustained, and this induced manufacturers to develop ever more powerful chips. Hitachi's joint venture with Texas Instruments to design a 16 megabit DRAM signified the technological and commercial implications of complex applications brought about by the need to enlarge research and development potential to maintain a foothold in international markets (*The Economist*, 3 February 1990).

The uncertainties of the market, and the frequent fluctuations in company fortunes, injected a complicating factor and a degree of uncertainty. This tended to destabilise the employment base of high technology companies, making them particularly susceptible to changes in consumer demand and tastes while requiring them to be sensitive to rapid and frequent advances in technology. In this way, high technology companies were vulnerable to developments which they themselves set in motion. Consequently, IBM was challenged by the development of 'clone' systems made by other companies and which performed as well as the originals.

69

Software companies outpaced each other month by month to produce complex programmes both for the personal computer market and industrial and business purposes.

Crisis in Property Development and Construction

Corporate investment in the property market provided another of the main impulses for the 1980s boom in development of urban areas and the general improvement of many prominent urban environments and city spaces. The property boom was experienced in commercial, industrial and leisure sectors and came to be seen by governments and the private sector as an ingredient in the overall regeneration of the cities. Development companies were able to capitalise on advances in the construction industry which allowed for 'fast-track' site development. New techniques of financing, such as the use of 'off-balance sheet debt' allowed companies to proceed with major speculative developments. Partnerships with city governments opened up new opportunities for development companies in locations to which it had hitherto been difficult to gain access. The internationalisation of the property industry gave companies a broader worldwide perspective on property markets and (at least in theory) more effective hedges against individual national market downturns.

However, a major problem was that the property market was highly susceptible to recession (Houlder, 1990). In the early 1990s, urban renewal based on the market-led thrust of property was exposed to the full force of the property and construction slump. This applied in the UK and the USA where development and construction companies suffered as property prices fell and companies ceased trading. City governments which had placed emphasis upon a dynamic property sector were faced with the consequences of placing so much faith in the private sector (Harding, 1990, p.98). Urban renewal was thus seen to be intimately linked to the fortunes of the domestic and international stock and money markets as development and construction companies were hit by falling returns and gloomy predictions about the future.

The Junk Bonds Crisis

In February 1990, Drexel Burnham Lambert filed for protection from its creditors under Chapter eleven of the US bankruptcy law after defaulting on a US$100 million loan. It had been one of the Wall Street 'stars' of the 1980s which had grown from relative obscurity to rival America's largest bond underwriters and most profitable financial institutions. The demise of Drexel Burnham Lambert was seen in some quarters of Wall Street as an almost inevitable end for an 'upstart', trading as it did in junk bonds and thriving on takeovers and leveraged buy-outs funded by 'funny money'. However, the Drexel case was symptomatic of the virtual collapse of the junk bond market. Junk bonds offered high rates of interest for offering high risk. In the late 1970s and 1980s, many US companies made use of the junk bond market to overcome the problems which they faced in raising capital through the banks. Amongst the largest of the junk investors were the savings and loan (S and L) institutions, or thrifts. These institutions provided funds for mortgages for millions of ordinary people along the lines of the British building societies.

S and L depositors were protected against the risks of the junk market by federal deposit insurance. They were thus willing to become involved in junk even though this often implied a low yield compromise. Drexel broadened its base by linking with S and L institutions, insurance companies and others to buy each other's bonds thus creating greater liquidity in the market. Later revelations gave rise to allegations that Drexel had achieved this by various manipulations of the securities market. Drexel's exit from the market, together with the ending of S and L junk holdings, eliminated the liquidity and increased the risks associated with junk bonds.

At the time when the economy was booming, the market was confident and companies over-extended themselves. Cash seemed to be readily available as lenders were willing to offer credit at relatively low rates of interest. The Drexel collapse happened at a time when the economy was yet to face recession, but it was a collapse which took place when interest rates remained low but where risks were increasing. Drexel thus, in effect, made too much money available as the market was changing and had too much at stake in its debt-financed buy-outs.

The Savings and Loan Crisis

Of even greater significance to the junk bond crisis was that concerning the S and L institutions themselves. In 1990, they were in serious trouble. As Martz et al. commented in May 1990:

> Like a huge storm in the financial stratosphere, the great savings and loan scandal has been raging for years without really touching most Americans. Rogues and swindlers paraded across the business pages and through the courts, accused of looting astronomical sums. Inept state and federal regulators fumbled to figure out what was going on and bumbling politicians resolutely looked the other way, all the while accepting millions in campaign contributions. Like the sums involved, the scandal was too big and abstract for most people to grasp.
>
> (Martz et al., 1990)

The scandal involved the loss and theft of over US$250 billion in government money; a sum which prompted a massive and costly federal government bailout which involved takeovers of insolvent S and L's and the sale of their assets. The S and Ls took advantage of market deregulation, weak official oversight and the trust of depositors by investing in high risk real estate speculations and 'hot money' operations. The S and Ls developed their activities in this climate by paying higher rates of interest and generating additional funds from their investments. The deregulation allowed the thrifts to broaden their investment portfolios and become involved in many high risk property speculations. This created a mountain of bad debt in the US financial system which threatened to impose a burden on taxpayers for years to come (Barclays Bank, 1990). It involved the S and Ls as well as the commercial banks, insurance companies and other financial institutions.

Small savers were among the victims as the thrifts collapsed. The 'easy years' of the early 1980s had been ones in which the restrictions on S and L lendings had been reduced. Risky investments seemed to offer the chance of quick profits despite the warning signs that the property market was overheating. For Martz et al., the money involved

> was stolen or squandered on playthings and development projects that made no sense: shopping centres in the desert, overbuilt condos, grandiose new cities like Keating's Estrella project outside Phoenix with almost nobody to live in it. Most of the money is unrecoverable, and many of the projects will have to be bulldozed.
>
> (Martz et al., 1990)

The political implications were no less significant than the economic consequences of the S and L issue. The allegations revealed a connection between the thrifts and politicians who had backed their cause during the boom years. Campaign funds had been raised by S and L contributions; individual politicians were implicated in various S and L deals and agreements which were questioned by the Senate Ethics Committee. This concerned those such as the five senators known as the 'Keating Five' who were alleged to have received substantial sums from Charles Keating, head of the Lincoln S and L. It was Charles Keating who was alleged to have given Arizona state politicians large sums in connection with his investment and development activities in the state. Martz et al. refer to the 'hired gun syndrome' where S and Ls tried to bribe politicians and officials to lobby on their behalf and work in their interests.

Bank Failures

The prediction that crisis would hit the banking sector came true. There was a direct connection between the fate of the S and Ls, the condition of the property market, the looming recession in 1990 and the condition of America's banks (Barclays Bank, 1990). As with the S and L crisis, the banking crisis affected both the rich and the urban poor. It also provided a clear indication of the nature of 'the market' in that it drew attention to the potential for the market to fluctuate between boom and recession or, in its worst case, boom and bust.

Even before the banking crisis, the Bush administration had plans for banking legislation following a review of deposit insurance mandated by Congress. The Secretary of the US Treasury, Nicholas Brady, linked this issue with the wider one concerning the structure of US banking and how to make the system more profitable and competitive. US banking had to shed its outdated and restrictive framework. Indeed, the restrictiveness of the system resulted in banks entering into higher risk forms of lending such as to developing countries, commercial property companies and take-over dealers. Many banks therefore became vulnerable to fluctuations in the property market and to defaulting Third World clients.

One illustration of the crisis facing the US banking system was seen in New England where the region's second largest bank, the Bank of New England (BNE) was declared insolvent in early 1991. Forty five Rhode Island banks were also closed following problems with property portfolios and allegations of corruption. The recession in the New England economy made the problems worse as business confidence slumped and consumers became more cautious. The 1980s confidence in the Boston financial sector had thus been badly eroded as the regional economy suffered and the service sector and high technology and defence industries were hit by the national economic downturn.

The speculative real estate boom of the 1980s had been particularly evident in Boston and its suburbs where property prices rose dramatically and the banks and thrifts eagerly financed overconfident borrowers and backed the plans of expansive property development companies. Many thrift institutions raised additional capital by going public and relied upon federal insurance to cover the expansion of their liabilities. The BNE allowed 37 per cent of its total loans to commercial real estate in 1989; a pattern which was indicative of the trend in New England banking as a whole. Following the crisis, it was left to the Federal Deposit Insurance Corporation to bail out failed banks and to the administration to pursue reforms of the system in response to a spate of banking failures right across the USA (Shafroth, 1991).

The US market was said to be a 'black hole' for foreign banks (Lascelles, 1991). The US$352 million deficit reported by NatWest Bancorp (the US arm of the National Westminster Bank) was one of the largest of a number of foreign bank losses in the USA. It was even larger than the Crocker National Bank loss in 1984, which up until 1991 had been regarded as 'the byword' in discussions about US banking losses. The NatWest losses were mainly attributed to poor conditions in the US market (whereas the Crocker problem had been ascribed to poor management), particularly those associated with the downturn in the property market and increased scrutiny by federal regulators.

Other foreign banks also incurred losses in the USA, most notably Marine Midland (the New York subsidiary of the Hongkong Bank) and the First New Hampshire (owned by the Bank of Ireland). As in the case of the NatWest, these banks all had commitments in New England and the north-east where they had developed links with local industry and expanded into real estate.

SOCIAL CRISIS

The expectations created in the 1980s, both in Britain and the USA, about the seemingly ever-expansive market were counter-balanced culturally and ideologically by the persistence of economic decline and social inequalities. The two sides of the market coin went together. In boom times, the visible 'successes' of the yuppie could be measured in terms of the failures of the less fortunate. The spending power of the employed worker could be contrasted with the lack of opportunity of the unemployed. The 1990s saw the ending of the confident myths engendered by expansion as market failures impacted upon ordinary people and the illusions of the 1980s were tempered by the realities of recession.

Impoverished urban communities represented the consequences of the *lack* of capitalist market dynamism. They were communities in which neither 'the market' nor public support had produced significant benefits. They were places where the recession in property and banking only restricted investment in renewal and rehabilitation. They were places where competition was defined in terms of a struggle for scarce resources between diverse and often conflicting social groups. Urban policy-makers in the USA and Britain implemented policies which only moderated the continued process of decline. The distressed districts of the cities therefore continued to suffer from declining economies, unemployment and poverty. Social tensions remained despite substantial investment in community development and small business programmes. In Britain, the disorders of 1981 and 1985 bore resemblance to the US disorders in the 1980s, most notably in Miami, and Washington DC in 1991.

However, to assert that there was a connection between the implementation of market-orientated policies and social polarisation poses problems. It is difficult to assess precisely how specific economic and social policies impacted upon the cities. There is always a time-lag between economic change, the implementation of policy and the manifestation of discernable outcomes. In the USA, policy measures associated with the administration's strategies on deficit reduction and tax reform had long-term implications for the condition of the cities, but the immediate consequences of particular policies were often left open to political speculation.

There has also been considerable controversy about the implications of the movement of people and companies away from central-city areas. It has been difficult to define the impact of these movements on the living standards and conditions of those who continue to live in the distressed districts and who suffer as the result of 'spatial mismatch' (Kain, 1968; Holzer, 1991).

Changes are complex, and social, economic and political impacts interdependent. For example, the national unemployment rate 'masks a wide range of local conditions' associated with poverty and social distress linked to the initial incidence of unemployment (Barnes and Penne, 1984, p.2). This includes the increase in high school drop-outs, welfare recipients, dislocated workers, homeless and 'de-institutionalised' people (Barnes and Penne, 1984, p.7).

Unemployment may also be one cause of increased racial prejudice and discrimination in the urban housing market (Mieszkowski, 1979). The 'spin-off' effects of interrelated social distress factors have frequently been referred to as the 'poverty cycle' in which related aspects of distress create the conditions for further problems which require the direct intervention of poverty agencies to help overcome them (The Boston Foundation, 1987a, 1987b). However, it is still difficult to quantify precisely how the conservative policies of the recent past positively or negatively impacted and to what extent social problems could be attributed to specific policy measures (Lublin, 1986; Heidenheimer et al., 1990). This is especially the case in US states where total employment levels actually rose at the same time as manufacturing employment declined (Federal Reserve Bank of Boston, 1986, p.12).

The Urban Institute Study

Attempts have been made to evaluate the impacts of policy changes by focusing upon the local level. A study made by the Washington DC based Urban Institute assesses the impact of the Reagan administration in the cities (Peterson and Lewis, 1986). The study looks at areas of policy which had the most direct influence in urban communities. The authors argue that the Reagan years marked a departure from previous attempts to implement urban policy initiatives. The nation's cities felt particularly vulnerable under Reagan's economic regime, with the older cities, in particular, fearing serious reductions in programme funding. Despite such fears, the outcome was by no means so bleak since the cities themselves, and Congress, acted to offset some of the worst potential effects of budget reductions.

Given this, the study measures the extent of some of the major structurally related problems affecting US cities. The analysis of the comparative condition of the cities is important in this respect. Feder and Hadley

75

(1986), writing in the same Urban Institute study, report the results of research conducted in twelve cities between 1982 and 1983, at the height of the debate about proposed cuts in social and health services. The findings, in assessing the affects of cutbacks upon hospital care for the urban poor, concentrate upon cities with high levels of poverty and provide a comparison between cities in terms of the following:

1 the city's total population;
2 the percentage of the city's population in poverty;
3 the percentage of unemployed;
4 the percentage of the population receiving Medicaid;
5 income per capita;
6 the percentage of beds in private hospitals.

The comparative statistics, based upon US government sources, provide a penetrating view of the relative levels of deprivation and wealth in each of the major cities studied by the authors. The differences in poverty levels between 1980 and 1982 provide one indication of the economic condition of the cities. Not all cities experienced the same economic changes or increases in poverty. As shown in Table 3.4, many cities performed badly during the early years under Reagan, while others managed roughly to maintain their position. In Boston, for example, the percentage of those officially in poverty rose during the period from 23.2 per cent to 33.4 per cent. In Cleveland, the increase was from 25.3 per cent to 39.0 per cent.

Table 3.4 Poverty in the hundred largest cities in the United States 1980–82

	Cities with little or no change in poverty (56 cities)		Cities with high increase in poverty (36 cities)	
	1980	1982	1980	1982
Percentage below poverty[1]	28.3	23.3	25.7	36.2
Percentage with Medicaid	11.6	9.4	15.2	18.3
Percentage change in poverty 1980–82	–	3.9	–	46.7

Source: Based on Feder and Hadley (1986, p.42)
Note: [1] Percentage with income below 125 per cent of the poverty income level.

The research links the incidence of poverty to the provision of health services to the poor. Feder and Hadley conclude that public funding is essential to maintain the volume of hospital care received by the poor and, that under conditions prevailing in the early 1980s, most private hospitals were not using private resources significantly to expand services committed to care for the poor.

The outcome, according to the authors, was related to the inadequacy of federal policy which had failed to reverse the general economic decline of the cities. The implication is that while public policies do not necessarily cause social problems as such, they can seriously compound existing ones particularly where they are conditioned by macro-trends in economic performance, population change and so on. This was an especially important consideration when set against the changes taking place in the wider economy which produced expansion in certain sectors of the economy while allowing others to decline. Under such circumstances, there needs to be a recognition of the socially divisive consequences of such change and the affects of change with respect to the creation of new 'pockets' of disadvantage and poverty in society.

International Comparisons of Poverty

Heidenheimer *et al.* (1990), referring to the findings published by Palmer *et al.* (1988) using OECD figures, compare the percentages of people in poverty in different countries. The figures are presented in terms of the official USA government definition of the 'poverty line' converted to other currencies using OECD purchasing power parities. The statistics show that the USA fares badly in terms of the incidence of poverty amongst western nations. Britain is not far behind (see Table 3.5). These are the two nations that have strongly stressed the importance of individual self-help and self-reliance free from high state subsidy:

> Many of the policy differences visible today are related to historical trends. To live in Germany or Sweden, which committed themselves early to contributory social insurance, is to be part of a comprehensive system of public transfers that may seem almost as routine as a paycheck. In countries that embraced income entitlements more hesitantly, such as Australia or the United States, the distinctions between those who have and who have not earned their benefits are sharper, and one's feelings as a recipient might be rather different.
> (Heidenheimer *et al.*, 1990, p.245)

This should not be taken to imply that generous social security arrangements are the sole or even main explanation for the figures in Table 3.5. Heidenheimer *et al.* suggest that there is an association between poverty levels and the nature of public policies in this field, but other factors such as levels of national income, the nature of wage bargaining and other governmental expenditures should be taken into account.

Enduring Poverty

This raises a question about the condition of the cities and the influence of public policies. Peterson (1986) provides an analysis of the relationship

Table 3.5 Percentage of persons in poverty 1979–82,
international

Country	All persons in poverty
United States	13
Australia	13
Britain	12
Norway	9
Germany (West)	8
Canada	7
Sweden	6
Switzerland	6

Source: Based on figures presented in Heidenheimer et. al. (1990), p.253

between federal policy and changes in the condition of the cities in the USA. His study concentrates upon urban policy change and the cyclical behaviour of the cities. His findings indicate that during the first Reagan term the cities fared better than expected by those who were worried about the effects of budget cuts upon local services. The cities were remarkably resilient to the recession when compared to the national economy as a whole, but that still failed to eliminate their underlying problems. Even so, the cities seemed to be less sensitive to cyclical fluctuations because city managers had become more expert in devising economic strategies which built-in contingencies to meet policy shocks and marked changes in the economic climate.

Even if the cities were resilient in the way described, this did not mean that they were successfully dealing with social polarisation. There was little improvement, even during the relatively high economic growth period of the mid- and late 1980s. For blacks there was little comfort despite the very small improvement in their overall situation as revealed in official statistics. Those who were officially defined as being 'in poverty' (i.e. below the officially defined poverty level) totalled 32.5 million in 1987 (13.5 per cent of the population) as compared with 35.3 million in 1983 (15.2 per cent of the population). In 1983, there were 9.9 million (35.7 per cent of all blacks) below the poverty line, whereas by 1987 the figure had only slightly fallen to 9.7 million (33.1 per cent of all blacks).

In 1983, there were 4.6 million Hispanics (28 per cent) below the poverty line compared with 5.5 million in 1987 (28.2 per cent). Significantly, there was also an overall increase in the total number of people of all races below the poverty line between 1986 and 1987 (up from 32.4 million to 32.5 million) despite a 0.1 fall in the percentage figure. Moreover, the 1987 figure compared unfavourably with the overall total for 1979 which stood at 'only' 26.1 million. In addition, in 1979, there were 36.6 million people in the USA below 125 per cent of the poverty level – in 1987 the figure was 43.5 million (1989 US Abstract of Statistics).

Not surprisingly, such statistics caused public concern about the lack of progress of the poor throughout the 1970s and into the 1980s which was underlined by the decline in real incomes of families with children between 1973 and 1984. Research published in 1986 for the Congressional Joint Economic Committee showed that between 1973 and 1984 there had been a reduction of 8.3 per cent in the mean real income of all families with children in the USA. There had been an improvement after 1982, but the level of income measured in this way in 1984 was still below that of 1970 (Rich, S., 1986).

More recent statistics, published by the US Bureau of the Census for 1989, showed that during the Reagan years median family income was barely above the level of 1973 when expressed in real terms. Significantly, figures published by the Bureau of Labor Statistics showed that average real earnings fell through the 1970s and 1980s so that they were only equal to the 1961 level by 1989. The prospect was one of a further decline as the US economy moved into recession in the early 1990s. Moreover, there was an increasingly uneven distribution of income between the poorest and richest families which had become particularly marked since the late 1970s. The distribution in 1989 was roughly what it was in 1947 (see *The Economist*, 10 November 1990).

Controversy surrounded the figures. Conservatives, together with government officials and the House Ways and Means Committee, pointed to the smaller size of US families as one explanation for the relatively low level of family incomes at a time of increasing national income. There were changes in the method used to deflate the figures made in 1983. Prior to that date there was a tendency for the statistics to overstate inflation because of the way housing costs were handled. It could also be argued that social welfare benefits and non-cash income had offset some of the effects of low incomes. However, these factors had to be taken into account alongside others which tended to overstate rises in living standards, namely the fact that the figures do not take regard of the tax burden on poor families or the affects of 'diseconomies' arising from smaller family units (especially where a woman is the head of a one-parent family or where there are high childcare costs).

CONCLUSIONS

These considerations, combined with the social problems associated with 'ghettoisation' and the expansion of the 'black economy' in poor urban communities, provided a clear indication that the plight of the poor was a central aspect of the crisis of the high technology, service-orientated, late-twentieth-century American urban scene. Experience during Reagan's second term confirmed the trend towards increased poverty, the widening of the gap between those below the poverty line and those enjoying the benefits of increasing per capita incomes and the growth of an 'underclass' of permanently impoverished citizens (see Chapter 4 for a comparison with the UK's underclass). Reagan's expected 'trickle-down' of wealth to the poor as the result of national economic growth had not substantially materialised. In this situation, state governments assumed a more determined role in initiating economic development as the cities expanded partnerships with the private sector (see Chapter 5).

However, there were costs involved. An analysis in *Fortune* in April 1988 indicated that the spending of some 80,000 municipalities, state and local governments in the USA had added around half a percentage point to annual growth in gross national product over the previous three years. During that period, local and state governments were spending substantial amounts to cover federal shortfalls to attempt to boost local growth, but by 1988 there was a wider recognition that constraint was needed. This meant that local projects such as road building, education construction and hospitals were being evaluated more stringently. Indeed, state and local government spending upon construction projects of this kind fell from a peak of over US$57 billion in 1987 to just under US$55 billion by 1988. However, the most dramatic change was in the condition of state and local government budgets.

State and local government budget problems were indicative of the change in the economic climate at the local level. The situation was not too serious in the short term since state and local personal tax revenues were predicted to increase despite the uncertainties which remained following the overhaul of the federal tax system in 1987. For those states which linked their tax deductions to the federal code, the initial result was to create a windfall. The loss of many federal tax deductions following reform meant higher taxable incomes which effectively enabled the states to raise more from their local taxpayers although some states compensated tax downwards.

There was little in the Reagan–Bush approach to 'decentralising' responsibility for urban policy which could promise much comfort for the those who maintained that interventionist federal programmes were the answer to reducing social and economic inequalities within and between cities. Decentralisation of responsibility implied that federal government was shifting the burden of expenditure to the state and local levels. Real federal grants to state and local governments declined by 37 per cent between 1980 and 1987. Many were targeted for local purposes and it was these which were a prime target for federal expenditure cuts. Other savings resulted from the consolidation of categorical grants into block grants and through the resultant simplification of local programme administration.

Nevertheless, the problems of the cities developed historically over a long period which pre-dated both 'Reaganism' and 'Thatcherism'. During the 1980s, there were important market-influenced structural changes in urban economies which continued to affect the policy commitments of

URBAN AMERICA: HIGH TECHNOLOGY AND POVERTY

governments. These developments raised important questions concerning the workings of the market economy. There was no guarantee that the 'social' side of the social market would be effectively catered for through economic growth. The market provided consumer goods and produced a good standard of living for those in work, but it did not 'raise all boats'. In the absence of positive interventionist social policies, the market perpetuated social and economic inequalities.

4

URBAN BRITAIN: ENTERPRISE AND THE UNDERCLASS

As in the USA, important technological changes have occurred which have impacted upon British cities. There have also been population shifts and regional imbalances which have produced inequalities between growing and declining districts and regions. As in the case of the USA, Britain should be examined within a wider international context.

Economic Growth under Thatcher

In 1979, the Conservative party stressed the deterioration in Britain's economic performance, record levels of inflation and poor labour relations under a Labour government. This, combined with increasing government borrowing and a low level of business morale, contributed to the electoral defeat of the Labour party and the introduction of the new Thatcher vision.

Under the Conservatives, Britain's inflation fell substantially. Between 1979 and 1985, the increase in consumer prices (excluding mortgage rate fluctuations) fell from 13 per cent to 5 per cent. A survey presented in the *Barclays Review* (1985) explained the fall in terms of a decline in world inflation and as following from a reduction in Britain's domestic inflationary pressures. Whereas between 1974 and 1979 Britain's average inflation rate had been 6 per cent higher than that of the OECD nations as a whole, between 1979 and 1985 the margin averaged only 0.7 per cent.

In the early 1980s Britain was faced with an increasing rate of unemployment. Unemployment, standing at around 3.25 million, at the end of 1986, presented the government with a political challenge as well as an economic one. This enabled opposition parties to point to the adverse affects of stringent monetary policy, constraints upon national and local government expenditure and cuts in local programmes affecting the unemployed, the poor and ethnic minorities.

Britain's GDP fell in 1980 as the economy went into recession, although the rise in GDP over the six years following 1979 was around 7.7 per cent. The early 1980s recession saw the near collapse of important manufacturing regions such as the Black Country in the Midlands and parts of the industrial north-east and north-west. There was hardly any shift in the share of GDP towards North Sea oil and gas, which used relatively little direct labour. The new high technology industries also prided themselves upon their labour efficiency. The 1985 Barclays Bank survey concluded that the growth in GDP and unemployment could be explained by reference to demographic and other factors. With a rise in the population of those of working age and an increase in the numbers of women seeking an additional source of income, it was difficult to reduce the overall level of unemployment especially at a time of an increasing level of output per person. Non-oil output per person rose by 1.7 per cent per annum after 1979, compared with 0.4 per cent per annum in the six years after 1973 (Barclays Review, 1985, p.53).

Britain's International Position

Despite the mid-1980s upturn, the poor economic condition of Britain revealed a process of restructuring in what was a comparatively weak economy in international terms. As far as direct USA/UK comparisons were concerned, the US economy was substantially larger than the British both in terms of its total GDP and as expressed in terms of per capita income. Overall living standards in Britain were lower when measured on a per capita basis and by purchasing power parity (PPP) which accounts for exchange rate differences between countries (see Table 4.1). However, this should not necessarily imply that the condition of the urban poor in the USA was substantially 'better' than that of the urban poor in Britain.

Table 4.1 International comparisons at purchasing power parity and current exchange rates 1990

Country	$'000s at PPP prices	Income per head at current prices $'000s, based on GNP
United States	21.8	21.8
Japan	17.5	23.9
Germany	16.3	24.2
France	15.5	21.5
United Kingdom	15.0	17.0
Italy	14.6	18.6

Source: *The Financial Times* and Datastream International 1991
Note: Figures show relative purchasing power, converting each country's GNP into US dollars using purchasing power parity exchange rates measured against identical 'baskets' of goods and services in each country. Income per head at current prices based on GNP is also shown.

THE ECONOMY OF URBAN CHANGE

As in the USA, developments in high technology produced radical changes in the way UK companies operated. Like America, the "urbanisation' of Britain's rural areas continued as companies established production, distribution and administrative facilities in areas away from the central business districts of the major cities. There was an expansion of the service sector of the economy which was particularly evident in financial services, advertising, the entertainments industry and leisure. Property development, fuelled by foreign investment, was a motive force behind urban regeneration and of central importance in the Conservative government's urban policy.

Britain's London Complex

An examination of economic change in Britain shows the importance of the specific national characteristics relating to the process of change in the cities and in the urbanisation of city regions. In Britain, London and the south-east provided the 'lead' to the rest of the country. In terms of per capita income, wealth creation and economic development the south-east was the jewel in the Thatcher government's crown. The south-east accounted for 11 per cent of the total land area of the UK (being second only in size to Scotland) and accommodated 30 per cent of the UK's population. The region's contribution to the UK gross domestic product increased from 33.8 per cent in 1977 to 36.1 per cent in 1987 with a per capita income 19 per cent above the national average (HMSO, 1989, pp.14–15).

The pace of economic development in London and its hinterland was faster than in other regions throughout the 1980s. London set the style in the arts, media and fashion and stood way above any provincial city in terms of its amenities, attractiveness to investors as a major financial centre, place to reside and as a convenient commercial gateway to Europe. Despite the decline of Britain as a world economic and political power, London was still a major international city. However, the international dimension to London's role contrasted strongly with provincial centres. Other cities, such as Birmingham and Glasgow, aspired to 'international status' as they strove to attract inward investment, develop financial services and promote cultural development, but their record was constrained by the magnetic pull of the capital city.

The pattern of change and balance of development between Britain's regions was therefore significantly different from that in the USA. The south-east and London provided a magnetic attraction for business. It engendered the persistence of the historical London-centredness with respect to politics and culture and effectively diminished the potential of provincial centres as counter-weights to the allure of the capital. However, there was a weakness even in the south-east's apparent strength.

The internationalisation of Britain's economy, partly through inward investment, was evident both in manufacturing and service industries. There was a greater reliance within key sectors, and in the economy as a whole, upon foreign-owned companies and an increasing dependence of domestic employment upon decisions taken in the boardrooms of non-British companies. During the 1980s, Britain actually reduced or eliminated its capacity to produce certain manufactured goods through domestic companies. The service and manufacturing sectors were thus extensively internationalised. This produced a vulnerable economic infrastructure as the investment decisions were taken abroad. Britain's major cities, particularly London, were increasingly subject to external pressures which threatened sustained economic growth. This was felt in the manufacturing and service industries as well as the property market which was influenced by the activities of foreign development companies and financial institutions (Moir, 1990).

A COMPETITIVE INTERNATIONAL SETTING

A measure of the relative decline of the British economy in world markets was the nation's share of world trade in manufacturing. The decline in Britain's share of world trade in manufactured goods fell from around 10 per cent in 1972 (by volume) to under 7 per cent (by volume) in 1988. Much of this decline was attributable to competition from the rapidly growing far-eastern economies which came to challenge Britain in a variety of manufacturing industries including the automobile and textile sectors. There was also intense competition from Europe and Japan.

There were signs that the decline had been temporarily arrested during the mid-1980s when there was a modest revival in the overall performance of the manufacturing sector and an improvement in manufacturing exports (Barclays Bank, 1989, p.5). Even so, Britain experienced a serious problem with an adverse balance of payments by the end of the 1980s which was partly the result of a large increase in imports of manufactured goods during the 1985–88 consumer 'boom'. There were also signs that Britain was facing increasing difficulties in the manufacturing sector as export prices moved above those of other West European economies.

Fears were expressed both in the Labour and the Conservative parties. Michael Heseltine expressed concerns about the uncompetitiveness of the British economy and called for a more concerted government commitment to Europe in preparation for the Single European Market in 1992 and a positive response to Britain's entry in 1990 to the Exchange Rate Mechanism (ERM) of the European Monetary System (EMS). Prior to his return to the Cabinet under John Major, Heseltine stressed the need for government support in key sectors of the economy. There were many areas in which Britain was seen to lag behind its European competitors, most notably in the provision of effective industrial training and in industrial research and development.

Heseltine (1989) concluded that a Conservative government should accept European cooperation and greater political integration. Germany and France were out-pacing Britain in a range of industrial, commercial and social fields. In technology, foreign governments were prepared to fund substantial innovative initiatives while the British government was reluctant to commit public resources to programmes that involved high levels of risk. In the financial services sector, Heseltine saw opportunities for British companies. In banking and insurance, cross-border arrangements enabled companies to benefit from a freer and wider market environment. The European Community had a population of 320 million compared with the USA's 220 million. This suggested an enormous potential for British companies.

REGIONAL DISPARITIES

Chapter 3 indicated that while there are market-influenced disparities between nations in terms of economic growth, patterns of investment and per capita income, there are also national regional disparities. As suggested above, in Britain, regional disparities have been exacerbated by the decline of traditional industries and the loss of trade to foreign competitors. They have also been influenced by changing market conditions and government policies affecting company location (see Chapter 2). International capital has tended to favour particular locations within the UK, particularly those seen by foreign capital to be potential growth or industrial areas supportive of major manufacturing operations requiring a ready supply of labour. In some regions this has created a heavy dependence of local and regional economies, especially in manufacturing, upon foreign companies. Figure 4.1 shows the gross value added in manufacturing attributable to foreign-owned companies in the UK.

The Liverpool Syndrome

The connection between urban distress and the worsening of Britain's relative position in the world economy was recognised by politicians in all parties and remained a theme in discussions about the future of the cities and their ability to become internationally competitive. Heseltine (1987) highlighted the decline of Britain's manufacturing infrastructure and the run-down experienced in particular regions outside London and by once great cities in those regions such as Liverpool.

Liverpool provided one of the most dramatic examples of the existence

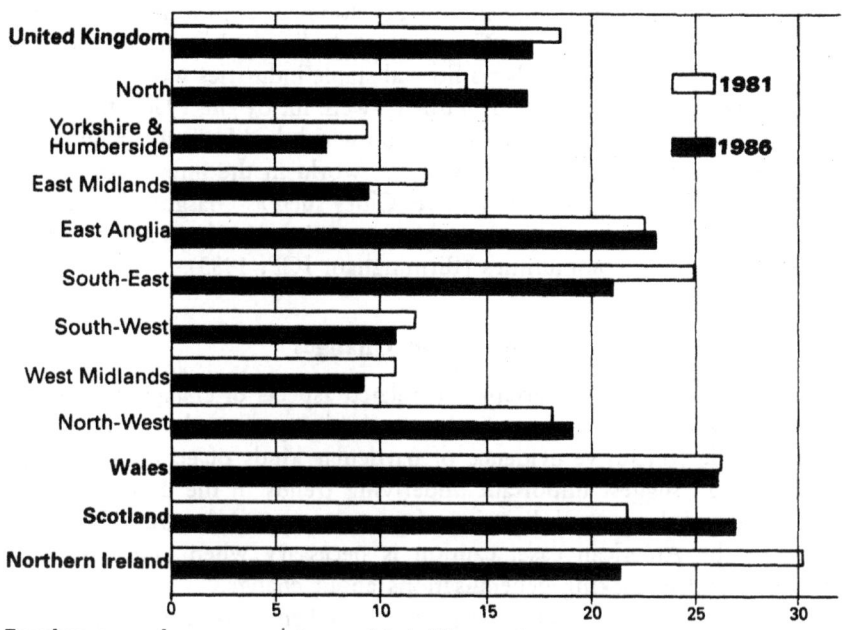

Figure 4.1 Foreign-owned enterprises classified to manufacturing: gross value added 1981 and 1986
Source: Regional Trends 24, 1989 (Business Statistics Office), reproduced with the permission of the Controller of HMSO
Note: percentage decline nationally is partly attributable to overall national economic growth in non-foreign owned manufacturing industries between 1981 and 1986.

of an urban crisis. The city was one of Britain's major trading ports between the early nineteenth and mid-twentieth centuries. The city's economy was undermined by structural economic problems and changes in methods of distribution and technology. Liverpool originally served as a major passage to the USA, but with the development of air freight and the introduction of new production processes, it could not remain internationally competitive. Britain's later orientation towards Europe, the concentration of regional financial services in nearby Manchester and the decline of Liverpool's traditional staple industries (including shipping and shipbuilding) helped to worsen the problems.

There were serious problems in other major British cities. The decline in the metal trades in the Black Country, to the north of Birmingham, left vast areas of dereliction in the 1970s and 1980s. The decline in manufacturing in Manchester, the run-down in shipbuilding on Tyneside and the closure of the docks in London were related to the changing position of

Britain in the world economy and the development of new high technology and service sector industries. The signs of decline had been evident as far back as the 1920s as Britain's traditional staple industries went into decline and as the provincial centres of industry were hit by the 1930s depression. In this context, the major cities of Britain which had been the manufacturing and commercial powerhouses of the world in the nineteenth century became internationally uncompetitive. Cities such as Glasgow, Manchester and Birmingham lost their 'international status' and suffered severe distress in their declining communities (Birmingham ICP, 1983).

Population Changes

As in the USA, two important and related aspects of change were population growth and redistribution. Redistribution patterns by themselves do not provide conclusive evidence of particular kinds of economic change, but they do suggest important underlying trends in the dynamic of the economy and the general direction of growth in particular areas. However, as with the American case, caution is necessary when examining such changes, and in making generalisations about change.

Regional differences are of importance in Britain because of the commonly acknowledged discrepancies between 'the south' and 'the north'. In reality, the pattern of regional imbalance is more complex than the popular notion of the north/south divide would suggest, although there are certain aspects of the caricature that have derived from observations of official statistics. There are also subjective perceptions which come into play when actual assessments of the physical conditions of 'northern' and 'southern' towns and cities are made. In very general terms, these are usually defined in terms of the environmental and physical decay to be found in cities in 'the north' and Midlands, and the relative 'prosperity' to be found in London, the south-east, East Anglia and the south-west (see Figure 4.2). Scotland and Wales are invariably consigned to the same category as 'the north' and Midlands. It is the areas of greatest decline that are therefore most readily associated in the popular view with the era of industrial expansion of the British Empire, and with its subsequent deindustrialisation.

In terms of population changes the picture is more complicated, although there is a discernable general trend towards population expansion in the southern counties. Champion (1989b) provides a summary of the pattern of population changes in Britain by taking a broad historical view of changes set against Britain's former role as the major world industrial nation. Urbanisation went together with industrialisation to such an extent that, by the middle of the nineteenth century, over half the population was living in urban settlements. By the middle of the twentieth century, there was a significant outward movement from the principal urban centres,

Figure 4.2 UK standard regions

leading to a net redistribution of population in favour of the smaller cities, towns and rural areas (Champion, 1989b, p.83). However, the urban–rural shift did not continue in one direction. More recent indications show that the loss of population from major cities slowed in the 1970s (with some cities gaining population) in spite of the evident continuation of overall population growth in non-metropolitan areas into the 1980s.

Champion highlights the cyclical nature of population deconcentration in Britain with its strongest acceleration in the 1960s and its subsequent waning. For Champion, this observation holds the prospect of relating trends in population distribution to periodicities in possible causal factors (Champion, 1989b, p.82). There have, for example, been changes in the pattern of employment opportunities as economic growth has affected different regions, towns and cities in an uneven manner.

Counterurbanisation?

As in the American case, the pattern of population changes raises the question as to whether there has been 'counterurbanisation' in Britain. Most of the research mentioned by Champion has been concerned with changes in population concentration even though not all studies have addressed the issue of counterurbanisation as such. Some have claimed that concentration around existing urban centres has been the most impor-tant factor in the British case, while others have emphasised the growth of smaller settlements and the rise in population in non-metropolitan areas as part of a counterurban process. Champion, however, uses what he terms an 'intermediate spatial scale' where 'counterurbanisation' in Britain is:

> interpreted as involving a shift towards a less concentrated pattern of population distribution, during which a negative relationship between the population size of a place and its rate of net migration change will generally prevail. A more specific definition requires that this shift reflects a switch in net migration balance rather than merely a change in the differential patterns of natural increase. It is likely that, at the lower end of the settlement hierarchy, this phenomenon will be accompanied by a rural population turnaround, in which areas which have for decades experienced net outward migration move to a positive migration balance. In general, if population distri-bution is shifting in favour of larger urban places, then urbanisation is deemed the dominant process, but if smaller places are generally growing faster than larger ones, then counterurbanisation is growing.
>
> (Champion, 1989b, p.84)

This is consistent with the definition provided in relation to the USA (see Chapter 3) although, as in the USA, the terminological use of 'counterurbanisation' may confuse the issue concerning the spread of urban forms of development into rural areas. As with the USA, it seems more appropriate to view the process as one involving 'urbanisation' in rural areas.

Moreover, there are comparative problems when a definition of 'place' is sought. It is important that the situation in Britain is, at least as far as possible, compared with that in the USA by using a satisfactory definition of similar types of 'place' (in terms of 'urban', non-metropolitan, etc.). Champion argues against the use of administratively defined local authority jurisdictions in the UK case. City boundaries in Britain simply do not reflect changes in suburban populations and fail to reveal 'the realities of the urban system' (Champion, 1989b, p.85).

As Champion points out, there have been various attempts to define urban areas in Britain in ways comparable to those used in the USA. However, problems with the available data necessitate the use of a combination of geographical frameworks to assess the nature of British population changes. The so-called 'functional regions' provide one of the most comprehensive frameworks. These divide Great Britain into a set of 280 local labour market areas (LLMAs) on the basis of 'travel to work' flows to employment centres. The approach identifies a set of twenty 'metropolitan regions' comprising groupings of LLMAs which have relatively strong links between themselves and which can be measured by using census data. This stands against other LLMAs which are relatively 'freestanding' (Champion, 1989b, p.85).

Champion concludes that the evidence for the existence of a counter-urbanisation pattern of population redistribution between 1971 and 1981 is overwhelming. Previous studies using standard local authority areas have pointed to the loss of population in the major cities during this period, and have shown that most parts of rural Britain maintained or increased their rates of population growth when compared to the 1960s. This was contrary to the national trend towards a lower level of overall population increase (Champion, 1989b, p.87). These observations have been confirmed by analyses based on the LLMA framework. The London LLMA, with 8.58 million people in 1971, lost around 8.6 per cent of its population over the next decade. A similar average rate of loss was evident in the LLMAs dominated by Birmingham, Glasgow, Manchester, Liverpool and Newcastle upon Tyne.

While the LLMA framework is useful, there are problems with providing a definition of individual travel to work areas. In some cases, the LLMAs are poorly defined. Neither do they define the parameters of 'city' change as such. If it is valuable to view 'conurbations' as a whole, then the data based on local authority jurisdictions may be useful (see Table 4.2).

It likely that the LLMA approach will be used to assess inter-regional population changes and economic exchanges by viewing British standard

Table 4.2 The populations of the largest cities and metropolitan areas in the UK

Individual cities based on local authority district data

City	1987 population
London (Inner and Outer)	6,772,000
Birmingham	998,000
Glasgow	716,000
Leeds	709,000
Sheffield	553,000
Liverpool	476,000
Bradford	463,000
Manchester	450,000
Edinburgh	439,000
Bristol	384,000

Metropolitan areas showing major city, based on local authority district data

Metro-area	1987 population
Greater London	6,772,000
West Midlands (Birmingham)	2,625,000
Greater Manchester	2,579,000
West Yorkshire (Leeds/Bradford)	2,054,000
Merseyside (Liverpool)	1,458,000
South Yorkshire (Sheffield)	1,296,000
Tyne and Wear (Newcastle)	1,136,000
Greater Glasgow	1,065,000
Avon (Bristol/Woodspring)	569,000
Cleveland (Middlesbrough)	555,000

Source: compiled from data in *Regional Trends*, 24, 1989, reproduced with the permission of the Controller of HMSO

regions like large LLMAs set within a European context. This would be especially appropriate when the south-east is opened up to continental Europe by the Channel Tunnel. It is an approach which is already implicit in the presentation of official UK government statistics on regional change (Clark *et al.*, 1986; HMSO, 1989).

'BOOM' AREAS

London-centredness has already been mentioned. London exerts a pull throughout Britain. Some commuters are known to travel hundreds of miles to the capital. Some workers stay in London during the working week and return to their provincial homes for the weekend. More significantly, there was an increase in the rate of internal migration in the UK which effected significant net gains for non-metropolitan counties, particularly in the south-west and East Anglia (OPCS, 1986).

London was not the only growth point in the 1980s. Despite the attraction of the south-east and its faster rate of expansion, there were important growth points and concentrations of wealth in the regions. The financial services sector expanded in Manchester, Birmingham, Leeds and Glasgow. Cardiff benefited from the influx of foreign investment in its hinterland. Bristol was regarded as a 1980s 'bright spot' as an increasing number of companies relocated their headquarters to the city.

The High-Tech Factor

The development of high technology industries illustrates the nature of the imbalance between locations experiencing expansion (particularly greenfield locations and the 1980s boom towns like Swindon and Cambridge) and the distressed areas in the cities which experienced the worst population losses, high unemployment and general decline (HMSO, 1989, p.43). As in the USA, companies involved in the development of the new technology tended to agglomerate where they could be close to other companies of a similar nature (Hall et al., 1987). The Thames valley to the west of London (including Berkshire, southern Buckinghamshire and parts of Oxfordshire) benefited from agglomeration to become a major concentration of multinational high technology industrial development. This was Britain's version of Silicon Valley.

Stuart Dalby (1990) suggests that the impetus for this came from the Thames valley's proximity to London's Heathrow Airport which gave company professionals in the computing business quick access to international capitals and commercial centres. Many electronics companies were subsidiaries of large American concerns which found Heathrow Airport to be a convenient point of entry for the rest of Britain and Europe. The Thames valley was close to London where senior foreign executives preferred to stay and conduct business rather than in Manchester or Glasgow. London provided the most important customer bases for the electronics industry. The Thames valley could easily be reached from Heathrow by the M4 motorway. Indeed, it was for this reason that the 'M4 corridor' between London and Bristol became a growth corridor during the 1980s.

Dalby alludes to how electronics manufacturers tended to be low volume, high value producers which did not export in bulk. Consequently, access to seaports was not an important factor – the new ports for the high technology industries were the airports. This was also an important factor for biotechnology and small electronic components companies which exported a large proportion of their products by air.

THE FRAGILE 'BOOM'

Despite the impressive records of the boom towns and centres of high technology, the 1990s witnessed the slowing effects of economic recession. The recession affected service industries as well as manufacturing and was felt even in towns like Cambridge and Swindon where high technology companies began to experience problems related to expansion and future investment. In the USA, the market approach to stimulating growth was marred by fluctuations on the stock exchange, the failure of the savings and loans institutions, crisis in the junk bond market and the problems of the banking system (see Chapter 3). In Britain, there were also serious weaknesses in the financial sector with banks facing a squeeze on profits. High interest rates were a constant source of complaint for companies and house buyers.

Up to 1990, the Conservative government could claim that new investment was creating employment in the depressed regions and that there was a new dynamism in the economy. Thatcher, like Reagan, had linked the growth of investment and the expansion of the wider economy to the prospects for urban regeneration and 'trickle-down'. However, both the inherent structural weakness and the internationalisation of key sectors of the economy raised doubts about this concept. If the underlying economic base was weak, how could growth and employment be sustained particularly at a time of high interest rates? It was a question that was crucial even in the boom towns of the Thatcher era.

The 1990s recession confirmed that the boom had been built on weak foundations. As unemployment climbed above the two million mark in 1991, industry again began to 'shake-out' labour. The south-east was especially hard-hit by an increase in unemployment in the service sector and unemployment in manufacturing rose nationally. Indeed, the rate of increase in unemployment tended to be highest in the areas which had grown fastest in the 1980s.

UK Manufacturing in Crisis

As mentioned above, one of the most critical problems remained the poor state of British manufacturing industry. The plight of manufacturing was one critical factor in the growth equation. An important specially commissioned report published in *The Financial Times* in November 1988 gave an early indication of the problems that would be faced in an economic downturn. The report doubted the underlying strength of the recovery which had taken place following the depths of the 1981 recession. It questioned whether the 1980s increase in manufacturing output was simply a response to a cyclical upturn in world markets, or representative of a more fundamental change in Britain's industrial structure.

The report surveyed thirty of the UK's largest exporting manufacturers. Over half reported an upswing in capital spending during 1988, but very little of this was going on the creation of new capacity (that is, on more factories or additional production lines). Instead, the expenditure was concentrated upon new equipment to reduce costs. Despite the strength of domestic demand and the large surge in imports to Britain in 1988, most companies reported that they were not losing sales through shortages of production capacity. This might be expected during a period of strong demand for manufactured goods, but as the survey suggested, British companies no longer made many of the products necessary to combat imports in some sectors. In addition, manufacturers were often cautious about expansion following their experiences during the 1981 recession. Also, the trend towards specialisation had produced less emphasis on volume *per se*, and therefore less pressure on factory space.

Demand was found to vary from industry to industry despite the favourable climate overall. Tobacco manufacturers were facing falling domestic sales and were shifting their concentration to Third World markets. Demand was stagnant in the drinks industry, while the situation in textiles was characterised by a steep decline in sales caused by foreign competition. However, in chemicals, pharmaceuticals, engineering, aerospace and electronics market conditions were more favourable.

With regard to international competitiveness, British companies were found to be 'by and large' holding their own in export markets. Manufacturers had not capitalised as much as foreign importers on the rise in domestic demand, and the impact of this upon the market share of British companies remained a potential problem. This was particularly evident where there was no longer any domestic-based production of particular goods.

Two problem areas were production machinery and small screen televisions. Many types of production equipment were found to be no longer available in Britain, while small screen televisions came largely from Japan, Hong Kong and China. Assembly plants were established by companies such as Sony in South Wales which used overseas sourcing extensively. The boom towns had courted Japanese and other foreign assemblers. However, the consequences for British companies were rarely taken into account when local authorities were keen to announce 'big-name catches' for their localities. Even in activities where British companies had chosen to concentrate, companies were reported to be 'defending their position'.

In 1988, labour shortages were beginning to be experienced. The main problems were in high technology and computer software companies located in the south, and there were problems for some process engineers. In general the companies surveyed by *The Financial Times* did not share the government's worry about labour costs, since most were expecting to hold down the level of wage increases in the coming year. However, information technology skill shortages were beginning to act as a constraint on expansion as larger companies sought electronics engineers and software specialists to update their operations. Many executives saw this problem as crucial in a competitive manufacturing environment where technological applications were becoming increasingly sophisticated and instrumental in promoting company growth. It was important that Britain's high-tech growth towns maintained their ability to deliver the goods, otherwise business would go abroad. It was therefore seen to be important to continue to increase productivity. The overwhelming majority of companies were convinced that productivity could be increased better to compete with foreign rivals.

International competition was also felt to be intense when it came to component sourcing and the importation of foreign component parts. Major companies reported that they now made less of their final product in Britain than they did a decade previously. This was partly due to the increased availability of cheap components in the Far East. In this connection, *The Financial Times* mentioned Cummins, the USA diesel engine company, which imported the core of some of its British assembled engines from the USA. Caterpillar, the earth-moving company, had closed two of its plants in Britain to source its products from France and Belgium. The APV processing equipment company stated that it was sourcing more components for its printing machinery from the USA as a currency hedge, while other companies indicated that they had deliberately dropped certain components where imports were the only remaining source. Such strategies contrasted strongly with West Germany and Italy which had both managed to retain more of their indigenous component supply industries intact. In addition, many of the production facilities in Britain were found to be foreign-owned and importers of components from overseas manufacturing plants serving British-based installations (see Figure 4.1).

The report stated one effect of these changes was to strengthen the link between final output and imports. UK-based producers could only make more by importing more, hence a rise in output automatically led to a greater import surge than in the past. In 1991, the House of Lords Select Committee on Science and Technology reaffirmed the importance of manufacturing industry to the future prosperity of the UK. The Committee commented that the UK's

> manufacturing base is dangerously small; to achieve adequate growth from such a small base will be difficult. A vicious circle has been formed. The balance of trade deficit puts pressure on exchange rates thus contributing to high interest rates. High interest rates discourage investment and further weaken our manufacturing base.

Most importantly, if

> market forces alone are to determine the course of events it is conceivable that we will end up with no significant British-owned manufacturing industry in the United Kingdom.
>
> (HL, 1991, paras 1.1 and 9.73)

In recession, towns that had experienced growth saw foreign companies modify their investment policies. Projects were either cancelled or post-poned. Local authorities had little choice but to wait for the next upturn as international corporate decision-makers reassessed their UK market operations.

Property and Construction in Crisis

This was evident in the property industry where there were concerns about the state of the market in all major cities in the UK. High interest rates hindered low-income city dwellers entering the housing market. The retail sector suffered a setback during 1990 as consumers responded to the discipline of costly credit and the effects of inflation. The effect was a slowdown in activity, a fall in institutional property investment and reduced returns on retail investment (Catalano, 1991). Schemes for new shopping malls were postponed or cancelled, store outlets were closed or scaled-down as companies competed with each other for higher market share (Gilmour, 1990). This was in contrast to the second half of the 1980s when cities vied with one another to attract retail developers and the big 'high street names' into themed malls modelled on the latest North American retail and leisure concepts.

At the end of the 1980s, there seemed to be no limit to the 'bonanza'. The subsequent property and construction industry slump provided a clear indication of the way in which market-led growth could falter (Gleeds, 1991). The recession undermined the Thatcherite vision of a 'property-owning democracy' and even affected central London and the prestigious Canary Wharf scheme in London's Docklands which had been one of Thatcher's most acclaimed schemes.

The problems were exemplified by the experience of the Reichmann family's Canadian real estate giant, Olympia and York (O&Y). O&Y's property holdings included the World Financial Centre in New York, the First Canadian Place office towers in Toronto and substantial stakes in international property companies. O&Y were the developers of the massive office complex at Canary Wharf. In 1991, the Toronto-based Dominion Bond Rating Service (DRBS) downgraded the debt secured on two major O&Y buildings in downtown Toronto. Moody, the US credit agency, cut the rating on US$548 worth of long-term debt secured on New York property from A1 to BAA1 as downward pressures on office rents affected the market. Concentration on New York and Toronto had exposed the company to the downturn in financial services; Drexel Burnham having vacated 5, 000 square feet of office space after the collapse of junk bonds (see Chapter 3). In Toronto, prestige occupiers such as accountants Deloitte & Touche Inc. and insurance brokers Marsh and Mclennan were reported to have left Canadian Place for 'cheaper accommodation' (Mortished, 1991).

DRBS noted that US$1.2 billion had been wiped off the O&Y share portfolio between 1990 and 1991. Big losses had been seen in O&Y's holdings in Gulf Canada Resources (hit by falling oil prices) and in property companies such as Trizec, which owns the US retail property giant the Rouse Company, the crisis-ridden Campeau Corporation and Stanhope Properties (joint owner of the large Broadgate development in London). Canary Wharf had also been hit by the downturn in financial services, climbing office vacancy rates and lower rents.

The UK thus faced a property crisis that was both domestic and international. The international interests of the large property companies like O&Y meant that the reverberations of downturns in the market were felt in both directions across the Atlantic. The link between property and the international financial markets provided an added ingredient simultaneously affecting developers in London, New York, Chicago, Toronto and elsewhere.

Jenkins (1991) points to the cyclical nature of the property market in the UK and the relationship between developers and domestic banks. Jenkins traces the pattern of property market fluctuations back to the 1960s and identifies the increasing involvement of banks in providing loans to development companies. In the UK, there was also the increasing involvement of the pension funds in the market. The pension funds found that they could not use the property market for 'tactical asset allocation' since property was not such a liquid asset as equities and mixed income securities. Consequently, net investment by the pension funds in property fell after 1984 to become negative (disinvestment) in 1990.

Between 1977 and 1986, the pension funds produced nearly half of net investment in property by long-term investing institutions. After 1984, they progressively reduced their investments at the same time that the banks were increasing their lending to property companies. As Jenkins (1991, p.77) comments, 'as one long-term investor was getting out of the market, the short-term providers of funds were dramatically increasing their exposure'. Indeed, the pension funds were selling assets to property companies which were then provided as security to the banks for further loans.

Property companies were aided by a bull market in shares which further helped them to raise funds for development. Bank lending on property reached $37 billion in 1991 with overseas banks taking over a major market share; 43 per cent compared to 22 per cent in 1976. By the early 1990s, UK banks were becoming aware of the consequences of the boom years. Jenkins suggests that there was an increasing desire by the banks to encourage property companies to strengthen their longer term debt situations. The banks were likely to become more reluctant to lend to the sector as debt increased and as companies needed to refinance their portfolios. This was already affecting retail, industrial and commercial property developments.

Financial Crisis

The property slump provided a reminder of the penetration of the UK by foreign capital and the vulnerability of a market dependent on borrowing. As in manufacturing industry, the property market was both weak in recession and increasingly dependent upon international investment. There was an international banking recession with the US banks in crisis and top banks like Barclays and the Industrial Bank of Japan loosing their triple-A credit ratings after reporting worsening results. In Germany, the big banks held their dividends as the situation in the east affected profits. Fears of a global 'credit crunch' were voiced by Nicholas Brady, the US Treasury Secretary, in April 1991 and nations were urged to lower their interest rates and ease monetary conditions.

The recession in the economy had a serious affect on UK banks. For example, the Royal Bank of Scotland announced a 58 per cent fall in interim profits in May 1991 and Sir John Quinton, the chairman of Barclays, was reported as saying that conditions were the worst he had seen in his 38-year career. The 'big four' clearing banks only showed improved results in 1990 because the previous years' figures had been depressed by heavy provisioning against Third World debt. Five UK banks actually failed, the largest being the British and Commonwealth Merchant Bank. All the big clearing banks announced major job losses involving over 30,000 workers throughout the country. All this intensified competitiveness in the banking and wider financial sector, both domestically and across the world.

The competitive environment was thus evident in stock exchange dealing. Deane (1990) spoke of a 'pitiless year' ahead in the London securities business. Despite the loss of jobs in the City since the 1987 stock market 'crash', there was more to come. There were several factors which were likely to affect London's overall standing as an international financial centre:

1 the increasing competitiveness of banking;
2 increasing competition on the foreign exchange market especially from Japan, the USA, France and Germany; the City was being challenged as Europe's pre-eminent financial centre;
3 there were problems with London's ability to regulate its markets;
4 London had to keep ahead with stock market technology and fast dealing and had to be administratively efficient in order to fend off foreign competition.

URBAN DISTRESS

Such trends and fluctuations in market activity helped to exacerbate the problems faced by those in the distressed districts of the cities. They also impacted upon whole city economies as professional staff were laid-off and vital services curtailed in response to government policies and financial constraints. The manufacturing crisis had a direct impact. The restructuring in the British economy, and the reversion to recession was accompanied by the intensification of social disparities and unemployment both in the most distressed districts and in the former boom towns.

The property and financial crises had a direct impact. Prospective 'first-time' house buyers could not enter the market. The slowdown in property development made companies more cautious about developments in urban housing and environmental improvement. Large companies had their profits squeezed as interest rates remained high and investments were more risky. There was therefore an *adverse 'trickle-down'* from the tertiary sector of the economy to communities which constrained them in their efforts to revive. These problems compounded those that had historically been a feature of UK cities (CDP, 1977; Cooke, 1989).

Indicators of Distress

In the 1980s, one way of measuring differences between British cities in terms of their relative levels of deprivation was by reference to the Department of the Environment's (DoE) indicators of deprivation. Those referred to most were population change; percentage of aged population; mortality rate; ethnic origins; level of unemployment; over-crowding in housing; lack of amenities in home; and percentage of one-parent families. The indicators were frequently referred to in connection with the designation of areas for classification under the Urban Programme (UP). Conferment of status under the UP brought important financial benefits to towns and cities which were most in need of additional resources to combat urban deprivation. While the DoE classification process did not produce the most satisfactory selection of areas for UP assistance, the indicators were useful in gaining an impression of the degrees of distress in different areas.

Criticisms of the DoE approach were made by local authorities, particularly those which had been excluded from funding under the UP. Aggrieved councils were joined by chambers of commerce and local trade unions in calling for more effective ways of allocating UP resources (Stoke-on-Trent City Council, 1985).

In an attempt to discover more about the allocative process used by the DoE, Bentham claimed that although the DoE had publicly declared the indicators that it used in arriving at its classifications of UP assisted areas, there had been only limited public access to information about the weightings given to each indicator (Bentham, 1985, p.122). The application of particular weightings by the DoE could, according to Bentham, significantly affect the ranking of towns and cities in terms of their deprivation. This affected decisions concerning the awarding of UP status to areas.

Bentham examined the DoE's use of the indicators and the actual UP designations. From the information produced, he inferred the most likely weights applied by the DoE at that time.

The results of his analysis suggested that the DoE classifications were most sensitive to the types of conditions indicated by high unemployment, overcrowded housing and high concentrations of ethnic minorities. In that respect the DoE's classifications appeared to be based upon reasonable assumptions about what constituted the most pressing urban problems. However, there were anomalies in the classification process resulting in serious inconsistencies in decisions affecting cities which were allocated aid under the UP and discriminating against those which were not. One reason for this was that the original DoE classifications were made in the mid-1970s. The relative positions of districts had changed in the intervening period. Newly available census information only tended to magnify the anomalies as a result of changes between areas. Bentham suggested that there were political factors which affected the original designations, given that they were made under a Labour government. Indeed, the 1986 reclassification of UP authorities by the Conservatives may also have been influenced by such political factors.

Relative Distress in the Cities

How did cities perform in terms of the DoE indicators? Table 4.3 shows selected cities from the 'raw' data referred to by the DoE after the 1981

Table 4.3 Selected districts showing indicators using 1981 Census data (for areas designated under the Urban Programme), UK

District	1	2	3	4	5	6	7	8
Birmingham	− 8.3	1.06	15.2	6.7	14.6	6.0	5.2	15.2
Bradford	− 0.9	1.05	12.0	6.3	13.9	5.5	4.5	11.2
Coventry	− 6.8	0.98	15.2	6.2	13.2	5.6	3.5	9.6
Manchester	−17.5	1.13	16.8	8.5	16.8	5.7	7.4	7.9
Leeds	− 4.6	1.01	10.3	5.8	15.7	3.3	2.4	4.0
London								
Hackney	−18.1	0.92	15.3	10.6	15.3	9.0	11.9	28.0
Haringey	−15.2	0.93	10.7	7.8	13.7	6.0	12.0	29.8
Tower Hamlets	−13.8	1.04	15.5	8.0	16.2	10.0	6.1	20.3

Key: 1 Percentage population change 1971–81
2 Percentage death rate
3 Percentage unemployment
4 Percentage single parent households
5 Percentage pensioners living alone
6 Percentage overcrowded households
7 Percentage households lacking amenities
8 Percentage of population in New Commonwealth/Pakistani households
Source: based on 1981 Census data

Census. This is free from weighting (the data is only roughly comparable to that for American cities in Chapter 3). Table 4.3 does not rank cities in order of their levels of deprivation or show the full list of cities officially assessed. Clearly, as suggested by Bentham, there was a need for a refinement of this data if levels of deprivation were to be assessed in aggregate terms between localities and if a satisfactory ranking of areas was to be achieved. Table 4.4. shows similar 'raw' data updated in a later government *Regional Trends* survey. The data is useful, despite the controversy over weightings, in that it illustrates some of the major problems faced in the areas listed.

Table 4.4 Selected districts showing indicators using *Regional Trends* data 1989, UK

District	1	2	3	4	5	6	7	8
Birmingham	− 1.9	11.6	49.1	45.0	17.7	3.5	2,166	nd
Bradford	− 0.6	11.9	43.4	18.0	17.1	2.8	694	nd
Coventry	− 2.5	10.8	48.5	13.0	17.7	2.3	888	nd
Manchester	− 2.0	12.6	43.6	24.0	18.0	4.2	684	nd
Leeds	− 0.3	11.3	39.0	27.0	18.6	1.0	1,064	nd
London								
Hackney	2.1	10.5	48.6	13.0	15.6	2.5	155	nd
Haringey	− 5.3	10.5	38.1	10.0	15.5	2.5	131	246
Tower Hamlets	10.0	11.7	48.8	8.0	15.9	2.8	185	303

Key: 1 Percentage population change 1982–87
2 Deaths per thousand of population, 1987
3 Percentage long-term unemployed, January 1989 out of total unemployment for area
4 Lone parent households (in thousands) 1986
5 Percentage of population over pension age (women 60 years, men 65 years) 1987
6 Percentage of households lacking inside WC 1981
7 Gross Value Added in Manufacturing (£m) 1986 (rounded)
8 Male average gross weekly earnings (full time £) 1988, nd = no data
Source: based on data in *Regional Trends*, 24, 1989

The Greater London Council Criteria

Davies (1984) produced a report for the former Greater London Council (GLC) which presented the results of an attempt to allocate an index of deprivation to each local ward in Greater London. The objective was to arrive at a ranking of areas in terms of deprivation using 1981 Census figures, thus improving upon the DoE's approach which, it was argued, had proved to be unsatisfactory in matching resources to greatest need.

The GLC study was subjective in that the variables used to indicate deprivation were selective. Nor did the areas which the study covered provide a full picture of the extent of deprivation in London. The study did not claim to provide a definitive ranking of London's distressed districts (Davies, 1984, p.1), but it was expected to contribute to the debate about the way in which deprivation was measured.

The study described the analysis of 1981 Census statistics undertaken by the Population and Statistics Group of the GLC Intelligence Unit. The analysis was based on work initiated by the DoE's Inner Cities Directorate, but the GLC's data were presented on a ward-level basis, in contrast to the DoE's local authority-wide basis. The GLC indicators were designed to provide the percentage of:

1 economically active persons aged 16 minus those who were unemployed;
2 economically active or retired persons who were semi- or unskilled;
3 households in permanent accommodation that were overcrowded;
4 households in permanent accommodation that lacked exclusive use of two basic amenities;
5 households with dependent children that were single parent households;
6 pensioners who formed one person households;
7 residents in private households who lived in a household headed by a person born in the New Commonwealth.

In this list, items 3, 4 and 7 were used in the DoE analysis, while 1, 5 and 6 were refined versions of the DoE indicators. The method of refinement used by the GLC was similar to that used by the DoE insofar as the indicators were combined into a single index using a statistical method known as the 'Z-score' technique. Using this approach, variables could be amalgamated into a single index for each area so that areas could be compared more directly. The London Borough wards with the highest Z-scores were those which displayed the highest levels of deprivation. The worst incidences of deprivation were found in the London boroughs of Tower Hamlets, Hackney, Brent, Lambeth, and Kensington and Chelsea (in that 'affluent' borough's Golbourne ward). The single index approach used in the Z-score analysis had the advantage of presenting data about deprivation in an easily accessible manner, although subjective factors were still assumed to play an important role in the making of decisions about urban policy objectives.

Census data (and subsequent updates) provided only a 'snapshot' of certain important indicators of deprivation. It made the comparison of data between areas possible only in general terms. Moreover, statistics characterising particular areas indictated liitle about the problems faced by groups and individuals suffering the consequences of poverty. There was therefore a need to supplement official views of 'inner-city problems' with specific information based upon local qualitative assessments and using alternative measures of deprivation and poverty which were often viewed as less important by governments. For example, income levels and the incidence of poverty needed to be assessed in terms of the relative positions of social groups and the changing nature of poverty.

THE UNDERCLASS: BRITAIN AND THE USA

The fuller interpretation of available data, together with a qualitative evaluation of deprivation and poverty, facilitated the use of the concept of the 'underclass' as a way of drawing attention to the plight of the poor in the cities (Murray, 1984; Oppenheim, 1990). This views poverty in terms of the social implications of the trends that are discernable in the official statistics and also trends that may be viewed more effectively through an assessment of the implications of longer term changes that take place. Instead of looking at 'poverty' as a static concept, it recognises its developmental nature. Its extent, nature and spin-off effects are constantly changing. Changes in the economies of the cities and regions of the USA and the UK have accompanied a social flux affecting levels of income, employment, crime and so on.

The underclass concept also has the advantage of being relevant to a comparative study of countries and their national levels of poverty, especially where a direct comparison of national statistics is made difficult owing to differences in methods of computation between nations. The focus upon lifestyle characteristics and qualitative assessments of attitudes and social habits helps to produce a culturally rich picture of social change. The approach is also useful in highlighting the polarisation taking place between social groups – those who are benefiting and those who are dramatically losing out.

One of the most articulate advocates of the underclass approach is the American conservative and political scientist, Charles Murray (1984). Murray's controversial view of the American underclass is as a social group defined by its abject poverty and by its inability or even refusal to integrate with the rest of society. In a summary of his ideas applied to Britain in 1989, Murray compares the development of an underclass in the USA with what he sees as a similar development in the UK. In Britain, the most depressed sections of the poor are becoming increasingly alienated. For Murray

> The difference between the United States and Britain was that the United States reached the future first. During the last half of the sixties and throughout the seventies something strange and frightening was happening among poor people in the United States. Poor communities that had consisted mostly of hardworking folks began deteriorating, sometimes falling apart altogether.
>
> Drugs, crime, illegitimacy, homelessness, drop-out from the job market, drop-out from school, casual violence – all the measures that were available to social scientists showed large increases, focused in poor communities. As the eighties began, the growing population of 'the other kind of poor people' could no longer be ignored, and a label for them came into use. In the US, we began to call them the

underclass. . . . What had once been a small fraction of the American poor had become a sizable and worrisome population. An underclass existed, and none of the ordinary kinds of social policy solutions seemed able to stop its growth. . . . Britain does have an underclass, still largely out of sight and still smaller than the one in the United States. But it is growing rapidly. Within the next decade, it will probably become as large (proportionately) as the United States's underclass. It could easily become larger.

(Murray, 1989, pp.26–7)

Murray concentrates upon illegitimacy, violent crime and drop-out from the labour market as the most significant early warning signs of the development of the underclass. These factors are interconnected and relate to other aspects of deprivation and poverty. The prevalence of illegitimate births, for example, is drastically higher among the working class than the middle class where career prospects are better and there is an expectation that women of middle-class origins will progress in relatively good jobs. Illegitimacy can also signify the breakdown of family networks and the instability of domestic social relations in a neighbourhood.

Murray ends his comparison of the USA and Britain on a pessimistic note:

So if the United States has had so much more experience with a growing underclass, what can Britain learn from it? The sad answer is – not much. The central truth that the politicians in the United States are unwilling to face is our powerlessness to deal with an underclass once it exists. No matter how much money we spend on our cleverest social interventions, we don't know how to turn around the lives of teenagers who have grown up in an underclass culture.

(Murray, 1986, p.46)

Oppenheim (1990), writing for the Child Poverty Action Group (CPAG), cites the above item and confirms the fears expressed by Murray about the UK. Using official UK government statistics, Oppenheim illustrates the polarisation between those on low incomes and those who benefited from the 1980s era of economic growth. The study compares British statistics with those for western Europe and the USA. For Oppenheim, the 'gulf' separating poverty from affluence in the UK has been widening as the result of urban decline, a generous tax regime for the better-off and pressures by central government to hold down benefit payments. Between 1979 and 1987, the richest fifth of society saw its share of total household income (after taxes and cash benefits) rise from 40 to 45 per cent while the poorest saw its decline from 6.1 to 5.1 per cent. Such problems were compounded by the rise in unemployment since the early 1970s and the increase in the number of people in low income and temporary or part-time employment.

Oppenheim (1990, pp.21–2) uses the 'poverty line' as measured by the level of supplementary benefit paid by the government. By that measure, for the UK in 1987, a two-parent family with two children below 11 years of age were living in poverty if they had an income (after housing costs) of £70.15 per week. According to this criteria, 10, 200, 000 people were living in poverty in Britain in 1987 (19 per cent of the population). Official statistics for the UK thus indicated the relative worsening of the position of the lower income groups during the 1980s despite a period when real incomes for the population as a whole were tending to increase. Top income earners went ahead at a faster rate than those at ther lower end of the scale (HC, 1990, 1991).

CONCLUSIONS

Britain and the USA face similar problems as far as the creation of the underclass is concerned. However, there is a sharp division between those like Charles Murray who see the problems of the cities in terms of dependency and those like Oppenheim who concentrate upon the structure of the welfare system and the underlying problems associated with urban decline. The idea that welfare systems produce a dependency culture which in turn fosters apathy and further decline is one that fits the public choice view of the failure of liberal interventionism. There is a stress upon personal and moral considerations (Oppenheim, 1990, p.15) which are seen to be major contributors to the plight of the poor. For Oppenheim, poverty is less concerned with the personal failings of the poor and more to do with the structuring of poverty within a system that discriminates against them in terms of employment opportunities and their access to satisfactory local social provisions and housing.

Both the USA and Britain experienced the effects of market changes during boom and recession. Britain was particularly vulnerable to the decisions of international companies in all sectors of the economy. Despite the USA's increasing penetration by foreign capital and the dependence of many American communities on the investments of overseas companies, Britain was especially exposed during the 1990s to the vagaries of the international economy. The UK did not have a substantial domestically owned computer industry and its indigenous motor industry had been reduced to a rump.

There were indications that the UK compared unfavourably with its major competitors with respect to its commercial and industrial attitudes. British industry was said to suffer from conservatism, backward cultural and social attitudes, short-termism and a lack of governmental commitment to supporting research and development (HL, 1991, p.7). Britain's regional imbalance towards London continued to work against provincial centres and to magnify the effects of economic downturn in the regions. In contrast, the globalisation of US industry was not be taken as an indication of that nation's loss of control over its economic base (Levinson, 1991).

As indicated in Chapters 2 and 3, economic decline was conditioned by factors which were beyond the direct control of local governments. Technological and economic change dramatically affected both the declining and the growing sectors of the economy. In the UK and the USA, the policies of governments could impact upon the conditions faced in communities resulting from economic change and unemployment. However, local governments were forced to cutback in crucial areas of concern to poor people. Political issues were thus coloured by the need for local groups to battle for resources in a competitive environment.

5

LOCAL GOVERNMENT AND URBAN POLICY IN THE UNITED STATES

This chapter covers the relationship between local and central governments and the implications of change for urban policy in the USA. Chapter 6 deals with developments in Britain.

In both the USA and the UK in the 1980s and 1990s, sub-national governments were intimately bound up with the performance of the private financial sector as pressures were brought to bear by restrictive central governments. The local provision of services in the USA and Britain became more risky as councils were exposed to the traumas of the market and the fluctuations in money markets and the national economy (see Chapters 3 and 4), and as local provisions were privatised and sub-contracted to entrepreneurial profit-seeking companies.

In the UK, Conservative governments pursued policies which severely limited the powers of local government. While the tendency was towards greater central control over local government, there was also a move towards a more fragmented and privatised provision of services, which in some respects resembled the American approach of local service delivery (Hambleton, 1990, p.8). In the USA, the Reagan and Bush administrations emphasised the transfer of responsibilities to the states. This often had the effect of restricting local governments in developing their own initiatives and programmes.

Set within the broad sphere of intergovernmental relations, the Thatcher government reduced the power of local government in a way that was regarded as unacceptable to many Americans (Hambleton, 1990). The British approach involved the use of, what was to Americans, a heavy-handed attack on the autonomy of local authorities linked to a part centralisation of urban policy initiatives. This contrasted with the US federal government's partial withdrawal from local-level provision.

Despite the localist rhetoric, there were a number of policy changes in the USA which constrained the autonomy of state and local governments. There were measures both to devolve US domestic policy authority and administration (Peterson, 1984, p.217) and to restrict sub-national governments in their ability to raise revenues. In this way, federal government

108

influenced state and local governments both directly and indirectly by changing their operational and financial environments. This happened during a period in the early 1980s when the *centralisation* of the federal *budgetary control* process became 'the engine of the administration's reorientation of domestic policy' (Salamon and Abramson, 1984, p.43).

In the early 1990s, US state and local governments faced a variety of financial problems which many regarded as placing the local government system in a 'fiscal strait-jacket'. Simpson and Jung (1990) point to a variety of problems where counties were 'coping with crisis' and trying to meet the challenge of the 'future shock' of 'unprecedented fiscal stress' into the mid-1990s. The problems included the impact of inflation on services; exploding justice system costs; a crisis in the health care system; problems with local government tax bases; underpaid local government employees; increasing costs of infrastructure replacement; rapid escalation of long-term financial obligations; and problems in generating revenue through property taxes. Many states entered the 1991–92 fiscal year without having settled their budgets (including California, Illinois, Louisiana, Massachusetts and Maine). The recession forced tax increases and service cuts both at the state and local levels. In Connecticut, the state faced a budget crisis while the city of Bridgeport filed for bankruptcy under Chapter Nine of the US Bankruptcy Code. New York City's Mayor announced a 'doomsday budget' while other municipalities faced increasing problems with budget deficits and dwindling resources.

THE UNITED STATES: A FEDERAL SYSTEM

Although this list would sound familiar in the UK, the fundamental differences between the British and American systems are important. These are evident in the organisation of sub-national government where there are different organisational structures, processes and public philosophies concerning intergovernmental relations and how central and local governments should interact.

The US system has guarded against the over-centralisation of power and authority. One of the cornerstones of the US Constitution was the provision of checks and balances between the different branches of government to ensure that liberties were protected and the rights of states defended. The partial 'separation of powers' was enshrined in the Constitution as a way of protecting 'we the people' against despotism and the excesses of those who may abuse power. Although local governments (including city governments) were not mentioned in the Constitution, the powers of the states were specified and protected. The US federal system divided government between the national government and the state governments. The powers of the national government and the state governments were part of a system which guaranteed a republican form of government and the basic rights of the states. There were powers which were exercised concurrently by the national government and the state governments (taxation, regulation of elections, commerce, judicial functions and general welfare spending) and those which were the specific responsibility of the federal and state levels.

Complexity and Diversity in US Local Government

A distinctive feature of the US local government system is its diversity and complexity. The diversity in local practices ensures that a wide variety of municipalities are defined as 'cities'. These vary in terms of their social and demographic profiles and economic and spatial characteristics. Boston, Massachusetts, faces different problems to those of a small city such as Redding in California. Boston has a population of over 562,000 and is at the heart of the Greater Boston Metropolitan Area with a population in excess of 3.5 million people. Redding has a population of only 80,000 and is situated in one of the most picturesque parts of California, enjoying access to rivers and the open country.

Table 5.1 indicates the numbers and different types of US local government. The basic functions and structures of local governments are sanctioned by the state legislatures, although counties and cities can frame their own charters and retain their own powers over finance. Therefore, local governments enjoy varying degrees of autonomy and, according to local circumstances, are not simply regarded as subordinate to the states. This has given rise to what has been described as a 'marble cake' federalism characterised by a complex intermingling of powers and responsibilities between the levels of government (President's Commission, 1960).

Table 5.1 Number of governments in the US

Type of government	1987	1982	1977	1972
Total	83,237	81,831	79,913	78,269
US government	1	1	1	1
State governments	50	50	50	50
Local governments	83,186	81,780	79,862	78,218
County	3,042	3,041	3,042	3,044
Municipal	19,200	19,076	18,862	18,517
Township	16,691	16,734	16,822	16,991
School district	14,721	14,851	15,174	15,781
Special district	29,532	28,078	25,962	23,885

Source: Bureau of the Census, 1990

There are many points of contact between state and local governments. Agency agreements ensure that services are provided in a coordinated manner between the levels of government and that there are effective administrative arrangements to handle federal programmes. For example, in social welfare there are links between state and city governments which ensure the efficient administration of welfare programmes and the distribution of payments to people receiving public benefits.

County Government

The states perform many of their most important functions through the counties (including the provision of public health, welfare administration, law enforcement, justice administration, roads maintenance and other administrative functions). Counties, like cities, vary enormously in terms of their size and economic profiles.

California provides an interesting example of diversity and how local government is organised. San Bernardino County in California is the largest county by area in the USA with a territory of 20, 164 square miles. Los Angeles County, including its urban catchment, has a population of over eight million people, while other counties in the same state have under 50, 000.

Californian 'general-law' counties were established by the state legislature. They allow voters in the each of the counties to elect a board of supervisors, sheriff, district attorney, coroner, assessor, tax collector, treasurer, and other officials. The boards of supervisors act as the county chief legislative and administrative authorities and are empowered to appoint other less important officers.

In California's 'charter counties', the counties are allowed to frame and adopt their own charters which may be drafted by the board of supervisors or by a special charter commission consisting of citizens elected by the voters of the county. The charter is submitted to the voters for approval in a special election. This system allows for major changes to be made in the legal standing of counties whereby a general-law county may see it as being advantageous to become a charter county so as to enjoy an enhanced flexibility in determining its own internal arrangements. In California, the urban counties have tended to favour the charter county status (Hyink, Brown and Provost, 1989).

City Government

Cities, or municipalities, are in many ways like the counties. They are sanctioned by the state legislature and both act as administrative agents of the state. Inevitably, the policy choices of municipal governments are affected by the limitations imposed by their state governments especially where these involve public expenditure issues (Davies, 1980, p.19). However, the city governments operate with a greater degree of autonomy from the state than do the counties. Cities are incorporated jurisdictions within the larger counties and they provide services which neither the county nor the state provide solely within that jurisdiction. The cities have greater control over their own money and enjoy important economic, land use and planning powers.

There are around 450 incorporated cities in the state of California, with thirty new cities formed since the end of the 1970s. In some cases, communities have sought to be 'annexed' so that they may come under city jurisdictions. Annexations have been instigated by residents in communities who wish to enjoy the benefits of city status. The majority of Californian municipalities are 'general-law' cities, but as in the case of the counties, they can frame their own charters to become 'charter' cities with enhanced status and greater autonomy (at least, in theory) from the state.

San Francisco is one of the few city–county governments in the USA. The city and county of San Francisco has one board of supervisors overseeing local services. It consists of eleven members elected 'at large' (on a city-wide basis) for a four-year term. The mayor is elected 'at large' every four years.

The Mayor–Council Form of City Government

The traditional form of US city government has been one in which the city council and the mayor are elected separately by the voters of the city. However, this is a much less popular model than it used to be, and only one-eighth of California's cities can now be defined in terms of this 'mayor–council' form of government.

Within this category, there is a distinction between *strong-mayor* and *weak-mayor* cities. In the strong-mayor cities, the mayor is the principal administrative officer with the power to appoint many of the city's officials and usually with veto powers over the council. Such an approach has been characteristic of America's big urban governments. Los Angeles, New York, Chicago and San Francisco are examples of cities which have this form of government, although the powers of the mayor differ in each case. In Los Angeles, the appointive power of the mayor is more limited than in some other cities. Some major services in Los Angeles are operated by boards of commissioners appointed by the mayor, but with the consent of the city council. Commissioners serve longer terms than the mayor, and no board member may be removed by the mayor without the consent of the city council.

In the weak-mayor (or strong-council) cities, the city councils have substantial administrative and legislative powers. Here, the mayor is primarily a figurehead possessing little administrative authority. The mayor is usually a member of the council who is selected by council members to serve as chairperson and mayor for a one-year term. The city's administrative posts are filled either by election or through council appointments.

The mayor may assume certain administrative duties or be assisted by a city clerk. In some cases, mayors are elected by way of local 'at large' elections, but this does not affect their powers in the weak-mayor model.

Professionalisation and the Council-Manager Plan

During the 1960s, the profile of local government underwent an important series of changes which highlighted the variations in organisation and structure at the local level. Many cities, towns and counties began appointing officials who were responsible for overall administrative affairs and there were developments with respect to governing councils and regional coordination between local governments.

Research carried out by the International City Management Association (ICMA) provided information about the numbers, types and managerial approaches of recognised municipalities in the USA (ICMA, 1989). The research described how municipal management in the USA and Canada had its roots in the *council-manager plan*. By 1989 there were 2,722 cities and counties in the USA which operated council-manager systems. The ICMA recognised the development of what it termed 'general management municipalities' which were also based upon the notion of professional management at the local level.

The typical city manager supervises the administration of the city and has the power to appoint and remove most of the officials at the head of administrative departments. The manager is selected for an indefinite term and holds office as long as he/she has the support of the council. The mayor in such cities has mainly ceremonial duties to perform, as in the weak-mayor cities.

Interventionist Urban Policy

'Urban policy' provided an added dimension to arrangements at the local and state levels. The federalist and strongly democratic/localist tradition in USA government engendered a suspicion of centralist solutions to local problems. The distrust of centralism was evident in public choice theory and in the policy commitments of conservative and liberal politicians. However, neither localism nor the existence of democratic structures were enough by themselves to ensure economic progress or the full realisation of citizen rights.

In the 1960s and 1970s American cities were the arena for dramatic and destructive riots and social violence. In 1963, President Kennedy appointed Daniel Moynihan Assistant Secretary of Labor responsible for policy planning and research. Moynihan joined the task force which planned President Johnson's poverty programme in response to the difficult situation in the cities. Moynihan prepared a Labor Department report in

1965 entitled *The Negro Family: the Case for Action* which pointed to the plight of America's black population in the inner cities. The study influenced President Johnson's approach to urban policy by emphasising the urgent need for action in the ghettos. The Johnson administration thereby adopted a public interventionist approach to urban problems which relied heavily upon federal assistance to localities by way of support to various anti-poverty measures implemented at the local level.

The interventionist approach was restrained by the Nixon administration in the 1970s, but survived under President Carter who initiated his 'new urban policy'. The Carter initiative was marred by 'well publicised bureaucratic turf fights, false starts and delays', however,

> the process was one that the administration touted as uniquely successful – primarily because of the massive involvement of virtually the entire range of government agencies and the probing examination of existing federal programmes affecting the cities.
>
> (*Congressional Quarterly*, 1978, p.19)

Carter presided over the expansion of several major programmes in the field of community and human development with an increase in financial support for urban housing and social services. There were measures to encourage the states to improve urban planning and create 'livable cities'. There was also support for the construction of urban mass transit facilities and pedestrian transit systems. There was a commitment to support joint public–private developments. Policies were developed to encourage the improvement of employment opportunities and economic development by way of 'soft' public works projects. One contemporary commentary reflected that in contrast to the urban strife of the 1960s:

> quiet summers have prevailed in most American cities in this decade. Though scars of past destruction are still in evidence, some localities have shown signs of progress. Once dilapidated neighbourhoods in densely populated urban centres of the Northeast and Midwest are being refurbished, while new financial incentives and lowering crime rates have begun to attract businesses and needed tax dollars back to the cities.
>
> (*Congressional Quarterly*, 1978, p.5)

'America in Ruins'

Two years later, the perception of urban America dramatically changed. The critics of public intervention claimed that there was little at the start of the 1980s to show that the cities were overcoming their problems.

In 1981, the publication of Pat Choate and Susan Walter's study, *America in Ruins* provided a controversial review of the deteriorating

114

condition of America's infrastructure. The study pointed to the downward trend of public works investment by all levels of American government over the previous twenty years. Decline was everywhere evident in the basic fabric of the cities of America despite the supportive policies of federal government.

There were important implications in the study for state and local governments with respect to the efficiency of their investments and the way in which federal government should go about making resources available to lower tiers of government. The question of debt financing and the procedures of local funding was high on the agenda. Debt financing was a costly way of funding urban infrastructure improvements. There had to be a way of controlling expenditure on such items. As in Britain, the question of high cost infrastructural investment was one that was crucial in discussions about the finances of local government particularly as it was an issue tied to that of the increasing costs of capital items during a time of rising interest rates (Smerk, 1991). In Britain, the Thatcher government curtailed capital spending and encouraged the private sector to invest in health services, roads and education. In the USA, the Reagan strategy involved adjustments to the ways in which state and local governments could raise money for capital items as well as measures to shift the burden of spending away from the centre.

The Move Away from 'Urban Policy'

Reagan's intention was to release resources for investment and growth by reducing the burdens of taxation and subsidisation. His first round of domestic budget cuts concentrated upon public programmes which had been targeted as unproductive and wasteful of resources and which, in many cases, had been deemed to be politically unacceptable.

Peterson and Lewis (1986) claim that no sector of the economy felt more threatened than the nation's cities. According to their analysis, carried out for The Urban Institute, the cities were singled out for budget reductions as part of a general policy to reduce the inflationary effects of public subsidisation. This approach was so determined that Peterson and Lewis claim that the administration had effectively disavowed 'urban policy' in the sense of compensating the cities for their economic decline and acute social problems. By 1988, the federal government was investing more in foreign aid than in its own cities and towns (Shafroth, 1989, p.115).

Wolman (1990) examines the impact of Reagan's early cuts on urban-orientated policies. His research found that the initial impact of the reductions was felt into the mid-1980s in both urban and 'implicitly urban' programmes (i.e. programmes which were not directly targeted at the cities, but which affected them). Outlays for financial year (FY) 1982 federal grants-in-aid to state and local governments (excluding grants to individuals) fell by US$9.5 billion from the 1982 pre-Reagan policy baseline.

The States and Economic Stringency

The attempt to shift responsibilities to the states (under the so-called 'new federalism') initially proved to be less successful, but the policy indicated a commitment which stressed the need for local initiatives which would be free from federal interference. The scope of state-level action was already considerable. Programmes involving local or city governments with the states were substantial. Important areas covered actions regarding local revenues, personnel matters, procedures relating to competitive bidding and the ability of local units to deal with physical problems (especially infrastructural expansion or replacement) and programmes involving the distribution of federal funds from the states to local government.

The economic recession of 1981–82 tended to heighten the adverse effects of cutbacks and withdrawal of federal support, but by the mid-1980s the states and city local governments had adapted to changed circumstances, often by generating their own sources of revenue and by entering into partnerships with the private sector. Frank Shafroth (of The National League of Cities) in *The Municipal Year Book* for 1989, contends that there was a growing lack of federal leadership and initiative which gradually 'made states the instruments of government leadership in the nation' (Shafroth, 1989, p.115). This allowed leadership at the intergovernmental level to devolve to the state level at a time when municipal government remained relatively constant in size. Even so, the federal bureaucracy managed to grow as overall spending at the national level increased (notably on defence up to the mid-1980s) in spite of the administration's attempts to constrain domestic discretionary spending (see Tables 5.2 and 5.3 and Figure 5.1).

Shafroth argues that the original intentions of Reagan's 'new federalism' were undermined because the states found it difficult to develop new sources of revenue as the role of central government declined. Instead of greater revenue sources to meet increased responsibilities, the federal government cut assistance in programmes affecting distressed urban communities and interfered with state and local borrowing and revenues:

> Direct grant assistance declined at a 5.3 per cent annual rate between 1980 and 1989. At the same time, the volume of so-called private activity tax-exempt municipal bonds was cut fifty seven per cent between 1983 and 1988. The 1984 tax law restricted the rights of municipalities to engage in public–private partnerships. The 1986 Tax Reform Act removed the deductibility of municipal sales taxes and imposed corporate taxes and higher issuing costs on traditional public

116

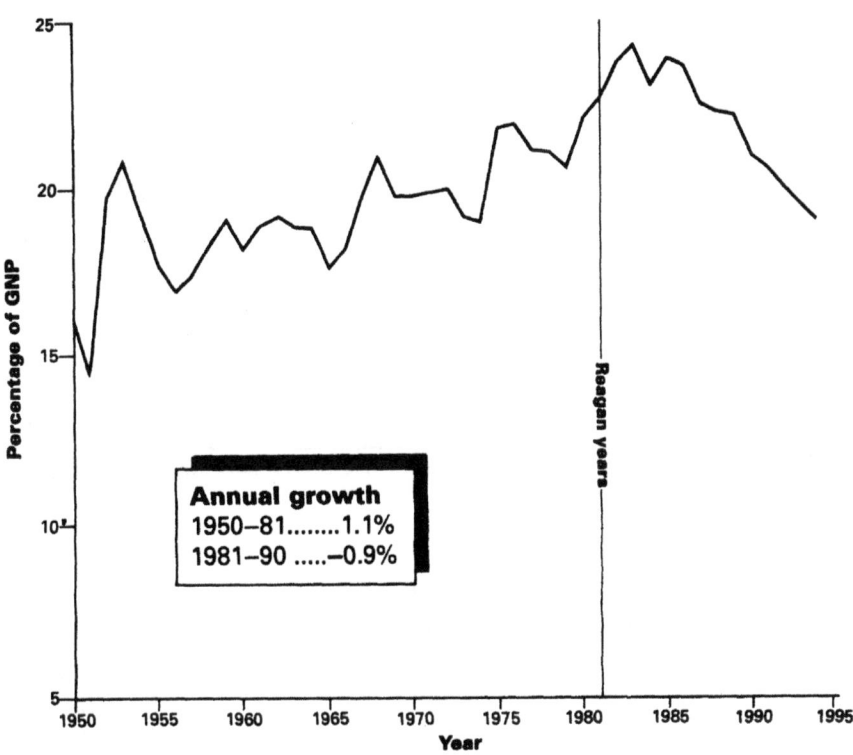

Figure 5.1 Federal spending in the USA: outlays as a percentage of GNP 1950–94
Source: Executive Office of the President, Office of Management and Budget, FY 1990, Major Policy Initiatives

purpose tax-exempt bonds . . . the federal government reduced investment in discretionary local programmes by 124 billion US Dollars
from the 1981 current service levels by 1988. The deepest cuts
occurred in programmes most affecting local governments: Medicaid,
education, aid for deprived children, public transportation, community development block grants, housing, municipal wastewater
construction grants, revenue sharing, health resources and job
training.

(Shafroth, 1989, pp.115–16)

Table 5.2 Federal government and national defence expenditure in the USA
1970–88

Item	Unit	1970	1980	1985	1987	1988
Budget receipts	$Bill	193	517	734	854	909
Individual income						
taxes	%	46.9	47.2	45.6	46.0	44.1
Income tax per capita	Dollar	441	1,072	1,398	1,609	1,629
Corporation income						
taxes	%	17.0	12.5	8.4	9.8	10.4
Budget outlays	$Bill	196	591	946	1,004	1,064
In constant (1982)						
dollars	$Bill	509	699	850	858	879
Defence	$Bill	82	134	253	282	290
Defence as % of						
GNP	%	8.3	5.0	6.4	6.4	6.1
Non-defence	$Bill	114	457	694	722	774
Deficit	$Bill	−2.8	−73.8	−212.3	−149.7	−155.1
Gross debt						
outstanding	$Bill	381	909	1,817	2,346	2,601

Source: US Department of Commerce Bureau of the Census, *Statistical Abstract of the
United States*, 1990
Note: Bill=billion

The Deficit Economy

Congressional concern over mounting national budget deficits and the
expansion of federal programmes led to determined action to reduce the
huge debt burden resulting from unbalanced annual budgets. Legislation
initiated by Senator Phil Gramm and his associates produced an anti-
deficit law which became known as the Gramm–Rudman law (the Balanced
Budget and Emergency Deficit Control Act 1985).

In February 1986, President Reagan submitted his fiscal 1987 budget
which appeared to meet the Gramm–Rudman objectives. However, the
budget greatly understated the level of defence expenditure and so
exceeded the new statute's target. Nor did the fiscal 1987 budget seem to
make any of the structural spending and revenue changes that were necessary to meet the law's reduced annual deficit projections. However, a first
round of cuts came into effect in March which were later affirmed by
Congress.

The Supreme Court found the original Gramm–Rudman automatic
spending cut mechanism to be unconstitutional. This removed the device
in the legislation which was intended to achieve a smooth set of budget
changes without recourse to 'across-the-board' cuts (such cuts were at
issue in subsequent budget debates about Gramm–Rudman, as in the case
of the FY 1990 budget where President Bush ordered across-the-board
cuts on numerous federal agencies following the failure of Congress to

Table 5.3 US federal outlays by function (billions of dollars)

Fiscal year	Total	National defence		International affairs	Health	Medicare	Income security	Social security	Net interest	Other
		Total	Department of Defence military	On-budget and off-budget outlays						
1976	371.8	89.6	87.9	6.4	15.7	15.8	60.8	73.9	26.7	82.8
1977	409.2	97.2	95.1	6.4	17.3	19.3	61.0	85.1	29.9	93.0
1978	458.7	104.5	102.3	7.5	18.5	22.8	61.5	93.9	35.4	114.7
1979	503.5	116.3	113.6	7.5	20.5	26.5	66.4	104.1	42.6	119.6
1980	590.9	134.0	130.9	12.7	23.2	32.1	86.5	118.5	52.5	131.4
1981	678.2	157.5	153.9	13.1	26.9	39.1	99.7	139.6	68.7	133.5
1982	745.7	185.3	180.7	12.3	27.4	46.6	107.7	156.0	85.0	125.4
1983	808.3	209.9	204.4	11.8	28.6	52.6	122.6	170.7	89.8	122.3
1984	851.8	227.4	220.9	15.9	30.4	57.5	112.7	178.2	111.1	118.6
1985	946.3	252.7	245.2	16.2	33.5	65.8	128.2	188.6	129.4	131.8
1986	990.3	273.4	265.5	14.2	35.9	70.2	119.8	198.8	136.0	142.1
1987	1,003.8	282.0	274.0	11.6	40.0	75.1	123.3	207.4	138.6	125.9
1988	1,064.1	290.4	281.9	10.5	44.5	78.9	129.3	219.3	151.7	139.4
1989	1,144.1	303.6	294.9	9.6	48.4	85.0	136.0	232.5	169.2	159.8
1990	1,251.7	299.3	289.8	13.8	57.7	98.1	147.3	248.6	184.2	202.7
1991 (estimates)	1,409.6	298.9	287.5	17.0	71.2	104.4	173.2	269.0	197.0	278.9
1992 (estimates)	1,445.9	295.2	283.0	17.8	81.3	113.7	184.8	288.6	206.3	258.0

Source: Budget of the United States Government, Fiscal Year 1992

complete action on deficit reduction). As the Fall Senate elections of 1986 came closer, there was widespread reluctance to vote in favour of further cuts as envisioned by the statute's remaining enforcement provisions. The Senate therefore voted against more cuts after suggestions that the deficit target would in fact be met following a revenue 'windfall' from US tax reform and in the light of other budgetary changes. It was calculated that these measures would reduce the forecast deficit for fiscal 1987 to below US$144 billion, which was less than the Gramm–Rudman target.

Despite these moves in the Senate, many Congressional leaders feared that the fiscal 1987 deficit would be far higher. The national economy was showing signs of slowing-down, and there were likely to be unforeseen consequences arising from tax reform. Congress also voted to increase the federal debt ceiling to a staggering US$2.152 trillion. On 15 August 1986, the Senate added a one-year Gramm–Rudman 'fix' which reduced the debt limit to US$2.111 trillion as a holding operation until the September of that year. However, the House went against the 'fix' in favour of a budget compromise on the reduced target.

Despite the Supreme Court's decision, the effects of the anti-deficit sentiment induced by the legislation were clearly evident in the President's fiscal 1987 budget strategy. Departmental and agency budgets were targeted to meet the proposed deficit ceiling and budgetary baselines for fiscal 1986 (which were used in the drafting of the budget) reflected the projections of the legislation and the constraints imposed by proposed yearly deficit reductions. The deficit reduction measures thus provided both the President and Congress with a justification for expenditure constraint on a wide range of defence and non-defence programmes, including those covering social, educational and economic areas of spending (Merkowitz and Cooney, 1986). In 1991, the deficit was US$318.3 billion (5.7 per cent of GNP), but was officially expected to fall to around US$2.9 billion by 1995 thus causing less concern for economic decision-makers.

TAX REFORM AS 'URBAN POLICY'

In September 1986, Congress passed a tax reform bill which was the end product of more than two years' debate about the structure of the US taxation system. The legislation had the effect of reducing income tax rates for individuals and corporations (a move that some critics of the President claimed to run counter to deficit-reduction measures) and reducing the number of so-called 'tax breaks' or tax exemptions. The expectation was that over the five years from 1987 to 1991 an estimated US$120 billion of the income tax burden would be shifted from individuals to corporations even as the corporate share of total tax revenues was still expected to remain below the proportion that had been achieved during the whole of the period between 1940 and 1979.

The new law reduced the top individual tax rate from 50 to 28 per cent, taxing 85 per cent of all individuals at the bottom rate of 15 per cent. Corporations were cut from 40 to 34 per cent and the law seriously affected real estate tax shelters and eliminated the lower rate on capital gains. However, some tax breaks survived in the final version of the law including deductions for state and local income tax and property taxes. These local and state taxes were the subject of controversy since deduction for non-business state and local taxes was the oldest deduction in the federal tax system.

Presidential attempts to abolish exemptions rested upon the argument that state and local liabilities should be included in the federal tax base as they were payments made by residents in exchange for public goods and services. Opponents claimed that this ignored the role by played deductibility in the fiscal coordination between the various levels of government and that many taxes were redistributive, providing benefits to socially deprived members of the community.

A comprehensive repeal of state and local tax deductibility would have affected large numbers of taxpayers in every state, being particularly burdensome on middle-income tax payers. The proposal would have affected intergovernmental relationships and fiscal relations between different levels of government. Many argued that the change would have produced financial pressures that would have reduced services and the prospects for sustained economic expansion, especially in those areas severely affected by industrial decline and restructuring (Northeast-Midwest Institute, 1986).

Relief for the Urban Poor?

Despite the retention of the deductibility principle, the administration claimed that the tax legislation represented an important means for reducing the level of poverty in the nation. President Reagan's statement made on 22 October 1986 claimed that the USA would

> have the lowest marginal tax rates and the most modern tax code among major industrialised nations, one that encourages risk-taking, innovation, and that old American spirit of enterprise. We'll be refuelling the American growth economy with the kind of incentives that helped create record new businesses and nearly 11.7 million jobs in just forty-six months. Fair and simpler for most Americans, this is a tax code designed to take us into a future of technological invention and economic achievement, one that will keep America competitive and growing into the twenty-first century. . . . For all these reasons, this tax bill is . . . a revolution. Millions of working poor will be dropped from the tax rolls altogether, and families will get a long-overdue break with lower rates and an almost doubled personal exemption. We're going to make it economical to raise children again.

The confident expectation that the tax changes would bring about social changes was designed to counter the arguments of those who were fearful that community programmes and local groups would suffer. Despite the potential advantages of lower taxation on the poor, the tax reform measures implied a further constraint upon public funding of social programmes. The combination of tax reform with the limitation of yearly budget deficits could only be effectively achieved by sacrificing the expansion of community programmes and by cutting back aid to the poor through reduced programme funding and erosion of social security benefits.

Tax-Exempt Bonds and Local Government

Tax incentives had a central role in urban renewal, as states and local governments devised strategies to overcome budget constraints and cutbacks. As Rich suggests, 'one could argue that much of the Reagan urban policy is manifest not in its budget documents but rather in the special analyses to the budget which details tax expenditures' (Rich, M. 1986, p.15). These indicated that while the overall level of tax expenditures increased dramatically during the 1980s, much of the increase was due to incentives related to housing and real estate development. These increased from about US$2 billion in 1980 to nearly US$19 billion in 1987.

Rich shows that one of the fastest growing categories of tax expenditure during the 1980s was the credit for the exclusion of taxes on interest earned in investment in state and local bonds. The volume of tax-exempt bonds issued by state and local governments rose from just over US$30 billion in 1975 to nearly US$200 billion in 1985 (Rich M., 1986, p.17). As the volume of bonds rose, so did the diversity of uses to which resources derived from the increase was directed. Rich indicates the spin-off for schools, hospitals, streets, sewers and other infrastructure, and also for convention centres, housing, student loans, pollution control and commercial developments. The use of tax-exempt bonds thus accompanied a substantial decline in direct federal support to the cities. This meant that indirect federal assistance had effectively increased due largely to tax incentives. The attitude of the administration was such that this development was seen to be contrary to its objectives of reducing federal support in the cities and controlling the size of the national debt burden. Tax reform was therefore seen as a way of reducing overall tax exemptions and, at least partially, reducing levels of state and local funding from tax exemptions.

The president sought new restrictions on the capacity of cities and towns to issue tax-exempt bonds and notes, although the original administration proposals to eliminate all industrial development bonds and many traditional public purpose bonds was not embodied in the final legislation. As stated above, the 1984 tax law effectively restricted the rights of municipalities to engage in public–private partnerships while the 1986 Tax Reform

Act imposed corporate taxes and higher issuing costs on traditional public-purpose tax-exempt bonds. There were also changes made to the rules governing such bonds. Negotiations between the administration and Congress blunted the full impact of the original Presidential approach on tax exemptions in this sphere, but still the changes represented an important challenge to 'traditional' bond financing arrangements. For conservatives, the changes marked a period in which discipline had been introduced into an area of public financing which had appeared to be getting out of control. The changes were consistent with a more ordered and prudent attitude to local finance in spite of the tightening of fiscal control.

In this, there are similarities with the situation in Britain at the time. Here was a conservative administration in the USA attempting to regulate the spending and financing arrangements of local public authorities. Ironically, this itself required positive central government action and a degree of centralisation of *budgetary control* over localities. Conservative governments in both countries were assuming a proactive role in this regard even though, as in the USA, the policy orientation of the administration was to withdraw federal support from key urban programmes. The British government went further and could combine this with interventionist policies (the urban development corporations) and direct attacks on the autonomy of local government while the Reagan administration used the strategy as part of an attempt to loosen federal ties to urban programmes. Both governments' policies were seen as being means to achieve a longer-term marketisation of local provision rather than being policy ends in themselves. Both approaches were tempered by nesessary political compromises which modified their initial reformist objectives.

In Britain, the government's policies were seen as a direct threat to municipal socialism and as a way of strengthening consumer accountability. In the USA, the objective was defined less in terms of ideology and more with reference to the constitutional position of sub-national government and the rights of states and local governments historically. Therein lay a further problem for conservatives. Were market-orientated policies always conducive to the defense of constitutional rights? Was it inevitable that at times there could be a conflict between valued rights at the sub-national level and the practical consequences of radical-reformist conservative policy objectives? Perhaps this was a dilemma that would naturally emerge whenever 'ideological' positions were being promoted.

CHANGING INTERGOVERNMENTAL RELATIONS

In light of the US federal deficit, the federal government also turned more to federal mandates as a form of off-budget financing. One area where this was the case was in environmental policy:

The environment was but one example of increasingly intrusive mandates that forced expenditures of local tax revenues without any choice by local officials. Especially in the tax area, the Reagan administration brought significant expansions at the expense of local governments.

(Shafroth, 1989, p.116)

The 1986 tax bill ushered a major change in the federal system by imposing costly mandates which had to be seen together with the direct interference concerning the ability and authority of municipalities to raise their own revenues. As indicated above, it had originally been proposed that the tax reform bill would eliminate the deductibility of state and local taxes and make the interest on virtually all municipal bonds subject to federal taxation.

The Congressional attitude to the proposals largely defended the basic rights of states and local governments to conduct their essential activities free from federal taxation and interference. However, less than two years later, the US Supreme Court concurred in *South Carolina* v. *Baker* that no such rights were implicit in the Tenth Amendment. The Court maintained that municipalities and states had no constitutional right to issue bonds on which the interest was exempt from federal tax. Despite the fact that this did not immediately affect the status of such bonds, the ruling challenged the basic notion that each level of government in the USA should be free of taxation by another level of government with respect to property or essential services. It thus provided an opportunity for future congressional and federal governmental interference in the affairs of state and local governments.

Shafroth argues that this had to be set against the inroads into the traditional authority of municipalities as Congress and the administration struggled to find revenue from almost any source to pay for growing deficits. In a series of changes beginning in 1984, the federal government imposed increasing constraints on the ability and authority of municipalities to issue debt, while imposing limitations on the municipal market. This adversely affected the borrowing ability of local governments at the very time that the federal debt was increasing. (Table 5.4 shows the debt and expenditure situation between 1970 and 1988 in USA municipalities and states.) According to Shafroth, Reagan

cut the strings to states and local governments and tightened fiscal constraints on the federal government which made for eight years of muscle-flexing by the states.

As the federal government struggled to keep its ship afloat, states undertook foreign trade missions, created new directions in early childhood education and pioneered historic welfare reform legislation.

This new-found power and leadership created consequences, opportunities and dilemmas for municipalities. As federal revenues declined or were shifted through state capitols and as state revenues and initiatives became more important, municipalities re-evaluated their relationships.

(Shafroth, 1989, p.116)

Table 5.4 US state and local government revenue and expenditure 1970–88

State and local government	Unit	1970	1980	1985	1987	1988
General revenue, total[1]	$Bill	130.8	382.3	598.1	685.6	727.1
From own taxes	$Bill	86.8	223.5	350.4	405.1	435.7
Property	%	39.2	30.7	29.6	29.9	30.3
Sales and gross receipts	%	34.9	35.8	36.1	35.7	35.9
From Federal Government	%	16.7	21.7	17.7	16.8	16.2
Direct general expenditures, total	$Bill	131.3	367.3	551.9	653.6	702.2
Education	%	40.1	36.3	34.9	34.7	34.6
Public welfare	%	11.2	12.4	13.0	12.6	12.7
Health and hospitals	%	7.4	8.8	9.0	8.7	8.8
Debt outstanding (end of fiscal year)	$Bill	143.6	335.6	568.6	718.5	755.0
Government employment (Oct.)[2]	Mill	10.1	13.3	13.7	14.2	(na)

Source: Department of Commerce Bureau of the Census, Statistical Abstracts of the United States 1990

Notes: (na) Not available
[1] Includes inter-governmental revenue
[2] Full and part-time employees
Bill=billion, Mill=million

On the one hand, this implied greater cooperation. Municipalities recognised that the federal government appeared to have neither the will, the vision nor the resources to meet emerging needs and problems at the local level. Thus key issues concerning day care, drug abuse and homelessness, which were not traditionally areas of state involvement, became more important at that level and offered cities different revenue sources. On the other hand, changes meant that direct federal support to municipalities was reduced or in some cases eliminated (the abolition of General Revenue Sharing and Urban Development Action Grants for example). This subjected municipalities to greater state scrutiny where programmes were state funded or where there was state control over authority to issue tax-exempt bonds (under the private activity volume cap). There were thus new programmes for the cities, but where the federal assistance was now channelled through the state rather than directly to the municipality. Both the new federal homeless assistance and anti-drug abuse programmes sent federal funds to the states prior to their allocation to the municipalities.

These changes created new layers of bureaucracy and additional incurred costs in meeting local needs and solving local problems. The states had greater leverage without being required to compensate local governments for what municipalities were asked to relinquish or contribute. The Reagan years thus 'reasserted a state primacy with unclear consequences for the future' (Shafroth, 1989, p.117)

Political Constraints: The Bush Administration

The Bush administration inherited the legacy of changes in intergovernmental relations from the Reagan administration. Any prospect of a major change in policies affecting the cities under Bush was largely set within the bounds of incremental adjustments to existing commitments. As seen in the introduction to this book, Bush initially adopted a conciliatory attitude towards Congress and preferred coalition-style politics to confrontation. This had its advantages in that it enabled Congress and the President to work together in the legislative sphere, but it created a tendency for policies to be 'fudged' to achieve cooperation and agreement. It came at a time when there were serious problems with regard to the budget deficit and issues such as the mounting costs involved in rescuing the savings and loan industry.

With Democratic majorities in both the Senate and House of Representatives, a Republican president could be sure that Congress would be keenly aware of his commitment on taxation and his plea for the electorate to 'watch my lips' in making his tax promise. This did not prevent Democratic Congressmen from blocking the administration's policies on matters which they did not favour (such as the administration's proposal to reduce capital gains tax) or resisting attempts to cut public expenditure.

Riddell (1990a) describes the political setting within which Bush operated in the early part of his term. There was, according to Riddell, no sense of 'national crisis' during this period, but a marked sense of apprehension about America's position in the world, its competitive position against Japan and concern about the social problems facing the nation. This included deep anxiety about the state of the nation's education, drugs and urban crime. For Riddell, the President had shown only 'the most general nod of concern'. The administration's response was to get by with a 'fudge' on the budget deficit and a neglect of education and the environment, two important issues raised in Bush's election campaign. This was characteristic of a classic 'muddling-through' which came under pressure in 1990 as the economy slowed down and high interest rates helped to increase the estimated budget deficit for 1991.

All this led to speculation about the President's commitment on taxation. The announcement in May 1990 that the President was not attaching any preconditions to negotiations on the budget led commentators to speculate that there might be a change of attitude on taxes. This was the one central issue where there was still a clear demarcation between Republicans and

Democrats. However, the Democratic leadership in Congress was not willing to take the blame of responsibility for increasing taxation to reduce the budget deficit – a situation which provided a debilitating stalemate on the budget, with Bush appearing to shift responsibility onto Congress.

The situation was complicated by the invasion of Kuwait by Iraq in August 1990. The invasion prompted Congress and the electorate to reconsider the implications of proposed budget cuts in defence. The cost of the American operation in the Middle East itself made original defence estimates redundant, but in the longer term the invasion served to focus attention on the ultimate impact of the 'peace dividend' and the end of the Cold War. The impact of defence spending was thus cast in terms of a trade-off in relation to domestic spending within the overall budgetary situation influenced by the federal deficit. There was a stark choice between defence spending and spending on domestic programmes covering education, social provisions and so on, in spite of Congressional resistance to cuts.

The prospect, at the end of August 1990, was one of real reductions in some programme expenditures to reduce the budget deficit in 1991. Table 5.5 shows the position in 1991 as it affected city priority programmes. Figure 5.2 shows the overall position regarding federal grants to both state and local governments up to 1990. The table indicates the tight financial situation facing state and local governments as far as federal support was concerned. Figure 5.2 indicates the burden of 'payments to individuals' which are funds devoted to social benefits and administered by state and local governments. A major indicator of fiscal stress for local governments in the 1990s was the net costs of mandated benefits. There is evidence that there was a shift in this burden to rural counties in particular as

Table 5.5 Funding for city priority programmes, USA

Programmes	FY 90	FY 91	Proposed FY 92
	(in billions of US dollars)		
Community Development Block Grant (CDBG)	2.9	3.2	2.9
Municipal wastewater	1.9	2.1	1.9
Airport improvement	1.4	1.8	1.9
Job Training Partnership	3.9	4.1	4.1
Anti-drug grants	0.4	0.5	0.4
Head Start	1.4	2.1	2.1
Assisted housing	7.8	9.5	9.1
Public housing operating subsidies	1.8	2.1	2.2
Homeless assistance	0.4	0.4	0.5
Others	2.7	3.0	4.0
Total	24.6	28.8	29.1

Source: National League of Cities, 1991

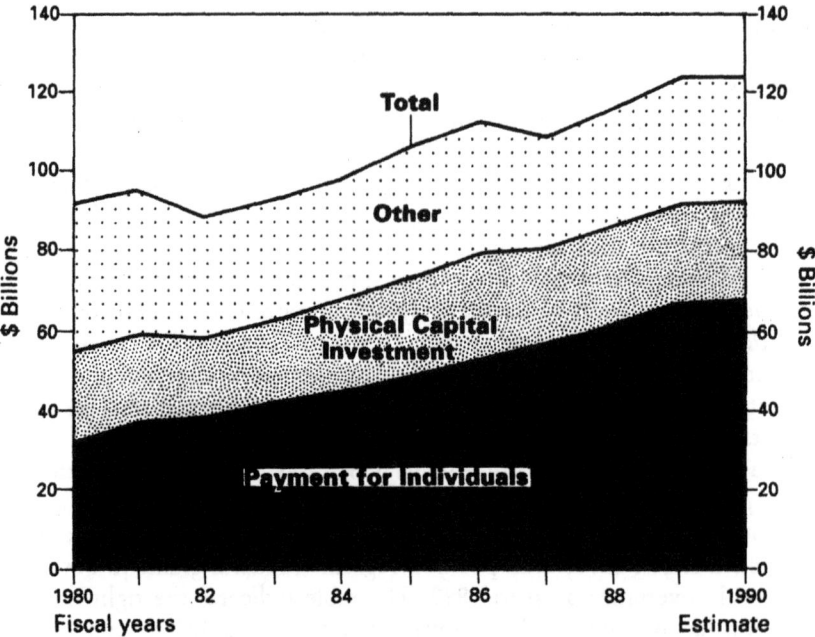

Figure 5.2 Federal grants to state and local governments: composition 1980–90
Source: US Government, *Special Analysis of the Budget of the United States Government,*
FY 1990

many low income families moved out of the cities to escape the high costs
of urban living (Simpson and Jung, 1990).

Bush's 'Turnback' Proposals

The Bush administration proposed a 'turnback' to the states of up to
US$22 billion. Potential turnbacks included Community Development
Block Grants (CDBGs), selected public and subsidised housing programmes and a range of education, environmental, justice, and health and
human service programmes. The states would take the funds as a block
grant. Many state governors feared that the turnback could be a new
version of the Reagan 'New Federalism' proposals to 'block' multiple
programmes as a preliminary to reducing their funding levels. This was
despite assurances by White House Chief of Staff, John Sananu that this
was not the basis of the proposal. Full programme funding would be
retained for at least five years.

Such assurances seemed to placate some governors, although many
remained sceptical about the President's claim that the plan would 'move

power and decision making closer to the people' (see Peirce, 1991). Congressional Democrats were also wary about the prospects of placing additional burdens upon the states and were critical of the administration's attempts to turn back programmes which had long been the subject of federal scrutiny as potential areas for cutback. There was particular criticism about the proposal to place the Community Development Block Grant with the states, since most of the prevailing grant money went directly to the cities. Indeed, this was seen as the last remaining 'string-free' money that went to the cities from federal government (Peirce, 1991).

CONCLUSIONS

Politicians and officials in US cities are less likely to expect central government assistance than their British counterparts. Despite the constraints upon federally funded programmes, the states and the cities have devised ways of generating funds both from the private sector and through the raising of local revenues. The prevalence of non-partisan agreements at the local level on priorities for development have assisted the process whereby cities have been able effectively to implement growth initiatives. The professionalism in local government, combined with this non-partisanship in county and city governments, presents a contrast to the highly politicised local government scene in Britain.

This is not to argue that non-partisanship and professionalism necessarily produce a more efficient system of city government in the USA. However, in contrast to the UK there is less of an 'ideological' conflict in local government. The USA does not experience the same conflict between the left and those supportive of private initiative. Although there is concern about the reduction of federal domestic programmes, American liberals are supportive of the market. The British 'new urban left' were highly suspicious of the enterprise culture in the early part of the 1980s. Their accommodation with the market was arrived at in the light of the successes and popularity of the market with the electorate and as Labour councils sought to generate economic activity by attracting private sector investment.

6

LOCAL GOVERNMENT AND URBAN POLICY IN BRITAIN

When compared to the USA, a study of British local government displays a different pattern of intergovernmental relations and contrasting expectations about the role and powers of local government. Britain's unitary form of government displays a stress upon the importance of a sovereign parliament, the strength of central institutions and the legitimacy of a government supported by the rule of law. This has made it difficult for advocates of local autonomy to make a case for strong local government (Jones and Stewart, 1985; Loughlin et al., 1985).

British Local Government After 1974

Apart from the existence of a range of quasi-governmental and regional bodies (Greenwood and Wilson, 1989), the British local government system is commonly described as a two-tier system. The lower tier authorities are the districts and the upper tier authorities are the counties (or regions in Scotland). There is a distinction between non-metropolitan districts and metropolitan districts. This is a legacy from the period when there were 'metropolitan counties' covering six of Britain's largest conurbations (West Midlands, South Yorkshire, West Yorkshire, Tyne and Wear, Greater Manchester and Merseyside). The metropolitan districts were a lower tier within the metro-counties. However, the metropolitan counties, together with the Greater London Council, were abolished in 1986 thus doing away with strategic city-wide governmental authorities in British cities.

There are many overlapping powers and shared responsibilities between elected local authorities. In England, the 296 non-metropolitan districts are responsible for planning, roads, housing, leisure services and economic development. The metropolitan districts have additional responsibilities for education and social services which, in effect, bestows upon them an enhanced status vis à vis the non-metro districts. The non-metropolitan authorities frequently complain about having to consult county authorities about such matters as planning, transportation and roads. Cities such as

Nottingham, Leicester, Preston, Stoke-on-Trent and Bristol feel that they have been adversely affected by the loss of their old county borough status following the local government reforms of 1974 which deprived them of the sort of powers traditionally associated with 'big city' status. Indeed, there was a real sense in which these cities were 'downgraded' in terms of their powers as a result of the 1974 changes.

Labour controlled cities have seen this as an additional infringement on local democracy, especially during the years of Conservative rule in Whitehall. Labour councils sought a return to city authorities with districts enjoying control over basic services. Many Labour councillors thus welcomed the Major government's commitment to review the structure of local government and the 1991 announcement by the Secretary of State for the Environment, Michael Heseltine, that the government favoured a review of the powers of local authorities and the abandonment of the two-tier system.

This indicated that there were serious problems associated with the prevailing system which was not conducive to effective local government in the cities, especially where there were conflicts of interest between city councils and non-metro counties with large rural constituencies. There were also problems associated with how the prevailing system would fit in with the increasing 'Europeanisation' of government where many policy issues have to be dealt with through new kinds of partnership between European, local, regional and national authorities (Rhodes, 1991).

The Political Setting

The Thatcher government's approach to controlling local authority expenditure involved the introduction of stringent controls over their spending powers, together with controls over the raising of local property taxes (the rates) and their eventual replacement (in England from 1990–91) with the controversial Community Charge (popularly known as 'the poll tax') imposed on virtually all people over the age of eighteen and a unified business rate which involved the setting of a rate for non-domestic property (NNDR) by central government. Local councils collected NNDR as agents of the centre; the money being paid into a central pool from which an amount was distributed according to a formula based on the proportion of Community Charge payers in the local authority's jurisdiction. Central government provided funds for local government through the revenue support grant (RSG) which replaced the former block grant after 1990–91.

Implementing the changes associated with central government spending restrictions and the poll tax proved to be a massive administrative task. Local councils had to invest in new computers, office space and administrative staff. There was also a need to monitor the collection of the tax and ensure that non-payment was kept to a minimum. Despite the attempts by many local authorities to offset the effects of central controls and fiscal restrictions, the government was largely successful in enforcing its objectives through a combination of persuasion and the initiation of legal actions against persistently non-cooperative authorities.

The 'privatisation' both of local services and urban policy was particularly controversial where local authorities were under Labour party control. By the mid-1980s, many Labour councils were dominated by left of centre council groups. Leftist local administrations were prone to be the highest spenders and the largest borrowers of finance for social, housing and other programmes, and it was this record of public support for local services that appeared to run counter to central efforts to control spending and borrowing at national level. The Conservative government's strategy was, therefore, to adopt measures and sanctions which could be used to force local councils into line with nationally set targets. Inevitably Thatcher's policies produced intense controversy and considerable conflict between central government and the localities.

Of particular importance over the period 1979 to 1991 were the following:

1 Thatcher's attempts to control local government expenditure and the replacement of domestic rates with the Community Charge and the 'nationalisation' of business rates.
2 The eventual proposal under the Major government to replace the Community Charge by a Council Tax. This heralded the reintroduction of a property tax, but (unlike the old rates) one which was combined with the remnants (from the Community Charge) of a 'head tax'.
3 The introduction of compulsory competitive tendering in selected council services and improved evaluation and quality assurance overall.
4 The introduction of radical educational reforms which transferred responsibility for major parts of the education system from local to national control.
5 The policy to 'break up' (Audit Commission, 1988, p.2) council estates and sell local authority owned dwellings to tenants.
6 A commitment by the Major government to review local government with the prospect of reforms to create unitary authorities to replace the two-tier structure and restore the status of cities which had lost powers under the 1974 reorganisation of local government.

The focus of legislation in these areas (with the partial exception of point 6) was directly aimed against the socialist notion of an 'interventionist' and proactive local government which the left claimed had formed part of a post-Second World War consensus on local administration. The challenge involved a move away from the idea that the public's needs for services such as education, community care and housing would be catered for by government. No longer would it be taken for granted that such services would be exclusively or mainly provided for from taxation or that local services should rest in the public sector.

High Risk Financing: From Cuts to the BCCI Collapse

Despite intentions to the contrary, the Thatcher government presided over an increasing level of overall public spending. However, the increase in money spending was deceptive. There was a decrease in the percentage of government spending in the total gross domestic product (see Table 6.1) during the Thatcher years and pressure on local authorities arising from inflation. The tendency was for local government spending to continue to rise in money terms but for services to be placed under stress as inflation eroded local budget allocations. The Major government projected an increase in spending while attempting to keep local government within the government's 'standard spending assessment' (see Table 6.2).

Table 6.1 UK general government expenditure (excluding privatisation proceeds) as a percentage of GDP[1]

1963–64	36.75	1964–65	36.50	1965–66	37.75
1966–67	39.50	1967–68	43.25	1968–69	41.50
1969–70	41.00	1970–71	41.25	1971–72	41.75
1972–73	41.75	1973–74	43.50	1974–75	48.75
1975–76	49.25	1976–77	46.75	1977–78	43.25
1978–79	44.00	1979–80	44.00	1980–81	46.50
1981–82	47.25	1982–83	47.50	1983–84	46.50
1984–85	46.75	1985–86	45.25	1986–87	44.00
1987–88	42.00	1988–89	39.50	1989–90	40.00
1990–91[2]	40.50	1991–92[2]	41.50	1992–93[2]	41.25
1993–94[2]	40.75	1994–95[2]	40.00		

Source: HM Government Budget Red Book, 1991, reproduced with the permission of the Controller of HMSO

Notes: [1] Based on money GDP figures adjusted for the years before 1990–91 to remove the distortion caused by the abolition of domestic rates.
[2] 1990–91: estimate; 1991–92: forecast; 1992–93 onwards: estimates.

Under Thatcher, many Labour local authorities strongly resisted central attempts to control local spending. Councils attempted to maintain services by borrowing more, taxing more heavily and using more imaginative ways of generating revenue through their investment portfolios such as the use of interest swaps deals with banks (which the House of Lords ruled as *ultra vires* following revelations about the debts run-up by a number of local councils, most notably Hammersmith and Fulham council), investment in higher interest bearing accounts and property and share holdings. The management of local authority investment funds became an issue of central importance as more local authorities made use of private sector fund managers and stockbrokers. In 1991, the collapse of the Bank of

Table 6.2 UK planning total and general government expenditure in real terms 1985–90 (figures adjusted to 1989–90 levels)

Item	*Out-turn (£ Billion)*				
	1985–86	*1986–87*	*1987–88*	*1988–89*	*1989–90*
Central government expenditure[1]	125.1	128.4	127.5	123.9	127.6
Central government support for local authorities[1]	38.8	40.2	40.8	38.9	38.2
Financing requirements of national industries	2.1	0.4	0.3	−0.5	1.1
Privatisation proceeds adjustment[1]	−3.4	−5.4	−5.9	−7.5	−4.2
Planning total	162.6	163.7	162.7	154.8	162.7
Local authority self-financed expenditure[1]	10.4	10.5	10.7	11.1	14.6
Central government debt interest	20.0	19.8	19.4	18.7	17.8
Accounting adjustments	3.8	3.9	4.4	6.3	3.7
General government expenditure	196.9	197.9	197.3	190.9	198.8

Source: The Autumn Statement (UK Government), 1990
Note: [1] Figures for local government spending relate to out-turns calculated in November 1990. Figures differ slightly from those given in Table 6.3 for 1989–90 in terms of out-turn. Figures are rounded and adjusted in line with government conventions contained in technical notes appended to the Autumn Statement.

Credit and Commerce International (BCCI), which held substantial local authority accounts, produced speculation about the future of some UK secondary banks (such as the Chancery Bank and the Edington Bank which collapsed while holding council funds) which were likely to suffer as councils became more cautious about their investment activities.

Arguably, the concern of councils to become more adept in what were often high risk market dealings was a consequence of the Thatcher era. Between 1979 and 1986 the government severely reduced resources for vital areas of local administration and service delivery at a time of price inflation, particularly in labour intensive services (such as education where the salary bill was a major element in service expenditure). The effects of the government's 'cuts' were felt both by service users and voluntary groups involved in various community programmes. The government's strategy initially rested upon the desire to reduce the level of public sector borrowing and the proportion of government spending in relation to the nation's national income (see Table 6.2), and this required the implementation of often drastic curtailments in spending for specific purposes (for example, on local government capital expenditure).

At the centre of the Thatcher government's attack on public provision was the attempt to control the apparently insatiable desire of public authorities to increase their spending, although the restraints on local government spending were 'less dramatic than some feared, or others hoped' (Audit Commission, 1988, p.3). This was partly because of the way in which local authorities devised ways to maintain their spending levels during a period in which central government was intent on curbing what it regarded as 'overspenders'. The various measures taken by central government to prevent the ever-resourceful local authorities finding loop-holes in government legislation and new ways of generating resources prompted what the Audit Commission described as 'a formidable system' of central regulations created to 'restrain the expenditure of local government in general, and high-spending councils in particular'. This involved:

1 tight controls on capital spending;
2 real reductions in grant and the introduction of associated penalties on high-spending authorities; and
3 the introduction of the Community Charge which was intended to 'sharpen the relationship between local taxation and local spending'
(Audit Commission, 1988, p.3)

The 'New Urban Left'

In the early 1980s, many Labour controlled councils regarded themselves as on 'the front line' in the defence of public services and jobs in the face of government policies which were claimed to be leading to a general decline in services and local economies. Labour councils felt during this period that a determined attempt to resist the Thatcher approach could lead to the government's abandonment of its restrictive policies (Jacobs, 1984).

The councils that adopted this view of local resistance have been variously described, but in general they have come to be seen as representative of the so-called 'urban left'. In practice, it was very difficult to define accurately a specific political characteristic which would produce a clear view of for what the urban left actually stood. The policies of different left-orientated councils differed widely and there was little consistency in the positions adopted by the various groups in and around the Labour party which adhered to left of centre policies. Stoker's definition well describes the urban left:

They are rather a diverse group incorporating younger Labour councillors, community activists and radical professional local authority workers who came to prominence in local government in the early 1980s. The origins of this emergence of local socialism can be traced back to the events and shifting ideas of the late 1960s.
(Stoker, 1988, p.193)

135

The early 1980s saw the conjuncture of a number of factors which, in many ways, 'forced' Labour councils into a defensive and activist position in spite of strong system constraints (Dunleavy, 1980b, Jacobs, 1984a). The urban left was also operating in a difficult environment. The run-down in manufacturing industry and the weakening of the trade unions, the increase in unemployment, the manifestation of 'leaderless' urban violence and riots and the coming to power of the Thatcher government all challenged the traditional ideological foundations of Labourism and created opportunities for Conservative councils to adopt innovative policies to implement the Thatcher vision of managerially efficient and cost-effective local administration (Holliday, 1990).

Opposition in the Cities

The opposition to the government was exemplified by the stand taken by the more vociferous Labour authorities in the metropolitan districts. This was initially in response to the introduction of a new system of central government funding under the 1980 Local Government, Planning and Land Act which replaced the old support system by a block grant. The grant was distributed on the basis of centrally determined grant-related expenditure assessments (GREs) which were to be calculated according to the assessment of local need as defined by central government. This was reinforced by the introduction of a penalty system for local authorities if they spent more than an average of 10 per cent above the needs assessment figure. There were also tighter controls over local authority capital spending by way of the introduction of cash limiting and tight monitoring of spending by the centre over services including education, social services, transportation, housing, etc.

The introduction of targets and penalties into the expenditure control system placed an added burden on councils. Targets were set in 1981 which involved the cutting of the block grant should spending levels exceed those laid down by the government (this was known as 'holdback'). As Stoker points out (1988 pp.156-7), the penalties were made increasingly severe each year so that by 1983-84 many councils were receiving less grant in absolute terms, at higher spending levels. The situation was made worse for councils by the constantly changing rules governing the targets and penalties which made it difficult for local authorities to plan ahead effectively.

The 1982 Local Government Finance Act retrospectively legalised the penalties and targets introduced in 1981-82 and dealt with supplementary rates and precepts by abolishing the right of local authorities to involve themselves in such activities. The legislation also set up the Audit Commission to oversee local authority finances and encourage councils to seek 'value for money' in their service provisions.

The Tide Against the Left

The ideological undermining of the mixed economy socialist approach of the Labour party at the national level took place at a time when socialist prescriptions were seen by the urban left at the local level to be especially relevant. However, in the early 1980s the political tide was flowing away from Labour in a way that was threatening the break-up of its electoral base in local government; a breaking of the public sector 'base' of labourism expressed in terms of public provision and municipal bureaucracy. Labour's ignominious defeat in the 1983 general election under the leadership of the veteran peace campaigner, Michael Foot, underlined the depth of the malaise faced by the party.

It was in this context that the new urban left came to assume a defensive position on the autonomy of local government and the commitment to the collective provision of local services. In so doing, the left discovered that it had to develop new ways to relate to the communities which it represented. Eventually, this led the urban left to reassess tactics and policies to the extent that, by the mid-1980s, there was a wider acceptance of the need to operate within the local government system and to make good use of a combination of litigation and imaginative approaches to financial management.

Defeating the Urban Left

The adaptation involved a compromise implying a move away from confrontation with the centre. It was brought about through the bitter experience of a series of legal actions, government penalties, political defections (notably to the Social Democrats who emerged as the result of a split from the Labour party in 1981) and electoral defeats for 'rebel' local councils which produced a more sober assessment of the tasks facing socialists in local government. From 1983, it was hastened by Labour leader Neil Kinnock's campaign to reduce the influence of the 'hard' socialist left within the party. This led to expulsions of left wing activists from the party and a major campaign directed against the Marxist Militant Tendency which had gained an influential position within some local party branches and on Labour councils.

The failure of the Militant Tendency to develop a substantial Marxist-style left alternative within the Labour party and the largely ineffective 'hard' and 'soft' left opposition to Thatcher's spending controls resulted in a re-evaluation of the role of activism at the local level. It also led to the establishment of new relations between the non-Militant left and the national Labour party leadership. This was achieved through intense debate within the Labour party about the role of the left and the meaning of socialism under changed political conditions.

The checking of militant council activism in the mid-1980s was one important factor in bringing about a changed working relationship between the left in local government and central government. In spite of this, Labour councillors who argued that major public support for local services was necessary to offset some of the effects of inner-urban decline were aware of the problems facing their communities, but were often seemingly unconcerned about some of the longer term effects of public subsidisation paid for with loans carrying high rates of interest and raised in localities which showed little prospect of generating sufficient economic activity to produce tax revenues proportionate to the expansion of local debt. Consequently, the government continued to pursue a restrictive policy, concentrating upon the adverse effects of high local taxation, capital spending and council borrowing (Birmingham City Council, 1990, 1991). There was also a desire by the centre to encourage innovation and efficiency in the provision of local services.

The Citizen's Charter

In 1988, the Audit Commission noted that the government had effected a significant change in the role of local government in Britain to the extent that there was no longer a clear consensus about local government's mission. Local authorities were no longer regarded as necessarily the only, or best, providers of their traditional services (a view of local government long prevalent in the United States). The implicit belief in the benefits of large-scale bureaucratic provision to assist with the solution of social problems had been replaced by a willingness to experiment with new methods and 'responsive structures' which would be more flexible and closer to the people. For the Commission, eight key factors stood out in the revolution that was taking place in local government.

1 Local authorities were finding it necessary to develop a better understanding of their 'customers'.
2 Local government would have to respond more effectively to the electorate.
3 Councils were encouraged to set and pursue consistent and achievable objectives.
4 Local authorities would have to continue to assign clear management responsibilities.
5 Councils would need to train and motivate people more effectively in the new era of managerial efficiency and cost-effectiveness.
6 There would have to be improved communications systems in local government.
7 Results would have to be monitored.
8 Councils would need to adapt quickly to change.

The essential prerequisite for meeting these objectives was that local authorities should continue to improve their management practices and develop ways of responding to their 'client groups' (Audit Commission, 1988, pp.1–2). The concern with services which would be responsive to 'customers' and which would be of high quality was one which was reflected in the Major government's idea of a Citizen's Charter (although the emphasis on quality was one that had been for a long time familiar to US state and local governments). This would enable the users of services in the public sector to obtain redress where service quality was below expectations (there were also to be provisions for monitoring aspects of the activities of recently privatised utilities). The common ground with the Labour party over the quality issue marked a change in emphasis for the Conservatives, away from radical policies to achieve consumer responsiveness in local government (such as the poll tax) towards a more 'caring' consumer-orientated approach (Stewart and Stoker, 1989; HMSO, 1991).

From Poll Tax to Council Tax

By 1991, the political setting had changed from the days of ascendant Thatcherism. The Labour party had reformed itself and now embraced the notion of 'opportunity' and the quest for quality in local government. In 1987, the party embraced what they had called at the time 'multiple purpose' unitary authorities and the abolition of the counties. Labour called for a revival of local government, not through grass roots militancy, but by way of reforms initiated by Parliament to give local authorities 'wider powers' (Labour Party, 1989, p.58).

The elements inherent in Thatcher's (and later Major's) competitive council strategy were embodied in a modified form in Labour's own programme. Labour had also gained in popularity as the poll tax became more unpopular. Serious public disturbances and the non-payment of the tax on a large scale prompted discussion within the Conservative party about the future of the tax. The Major government's difficulties over the question of what would replace the poll tax came at a time when Labour was more confident about its prospects for the coming general election.

Many 'Thatcherites' in the Conservative party regretted the abandon-ment of the poll tax and the measures taken by the government to offset its regressive tax effect. Some Tories were openly critical of the government for having compromised in the face of mounting opposition against the poll tax. One of the most obvious compromises involved the way in which the government handled the poll tax issue in the 1991 Budget and in subsequent statements on the future of local government financing. These moves culminated in the proposed return of a property-based council tax which was regarded by many Tories as an over-turn of the Thatcherite principle that each individual would pay towards council services so that they would directly feel the affects of their local council spending decisions.

The Thatcher government wanted to reduce the proportion of expenditure on local services financed by central government. Major's 1991 budget announcement enforced a £140 cut in Community Charge payments nationally by shifting the burden for local financing onto central government. The Community Charge was reduced, but this was achieved by raising the necessary additional revenue for local government by increasing the centrally imposed value added tax (VAT). The effects of this change can be seen by reference to Table 6.3. In addition, in 1991 the government announced that it would again 'charge-cap' selected councils under capping criteria announced in 1990. This included the capping of Conservative controlled Warwickshire County Council which was forced to cut a range of council services including education while many Labour councils adjusted their expenditure to avoid the capping threat (Vize, 1991a).

For 1992–93, the government projected a spending figure for local authorities which to many appeared to be flawed by the use of what Vize (1991b) called 'misleading' assumptions on inflation. According to the local authority associations, the government's under-estimation of inflation in looking at its Total Standard Spending (TSS) target (current and capital spending, financing costs, interest charges, etc.) seriously de-emphasised the pressures on services and the adverse affects on service quality especially where services were facing increased salary bills (fire, police, education and social work for instance).

Was the government's concern with tight spending regulations and a shift of funding to the centre conclusive evidence that the Conservative party had decisively become *the* party of centralisation? Certainly, the Major government's about-turn could not be presented as though it was part of any grand strategy. Major's top-down 'nationalisation' of local government financing was more commonly interpreted as a consequence of the political imperatives as seen by a government intent upon winning the next general election. However, as already indicated, there was a tradition of centralism in the British system of government which, combined with the Thatcher record, made it easier for a government to take such actions. The Labour party had at various times itself been responsible for centralising government departments, the expansion of public ownership and the control of local spending. Thatcher's urban policy record was imbued with a strong centralist element which, like the Community Charge, had been influenced by a desire politically to combat local labourism.

Table 6.3 UK general government expenditure showing estimates to 1995

	£ billion					
	1989–90	1990–91	1991–92	1992–93	1993–94	1994–95
Public expenditure planning total	162.9	180	205	221	232	243
Local authorities' self-financed expenditure	14.5	14.5	9	9.5	11.5	13
Central government debt interest	17.8	17.5	16.5	17	17.5	17.5
Accounting adjustments	4.2	3.5	4	5	5.5	6
General government expenditure[1]	199.4	216	235	252	266	279
Privatisation proceeds	4.2	5.5	5.5	5.5	5.5	5.5
General government expenditure taking account of privatisation proceeds	203.6	221	240	258	272	285

Source: HM Government Budget Red Book, 1991, reproduced with the permission of the Controller of HMSO
Note: [1] General government expenditure (including and excluding privatisation proceeds) and the public expenditure planning total are rounded to the nearest £1 billion from 1990–91 onwards; local authorities' self-financed expenditure, debt interest, accounting adjustments and privatisation proceeds are rounded to the nearest £0.5 billion. General government expenditure excluding privatisation proceeds is assumed to grow by 1.75 per cent in real terms in 1994–95.
General government expenditure includes debt interest payments to other sectors as follows (£ billion): 1989–90 18.5: 1990–91 18: 1991–92 17.5: 1992–93 17: 1993–94 18: 1994–95 18.

The Liverpool Syndrome

Despite the attempts to gain greater control at the centre, and the Labour leadership's own approach with respect to the control of its left, there were still indications that the issues of constraint and cutbacks were likely to be important issues into the 1990s. Charge-capped Labour councils felt aggrieved by what they regarded as the inequities of the limitations imposed by central government.

In Liverpool Labour experienced the embarrassment of a Militant-influenced opposition to government and party policies which drew attention to the tensions between the party leadership and those who desired an alternative to the party's moderate 'new realism'. There were special conditions pertaining in Liverpool, but despite these the role of the left and the issues relating to local service were relevant in other localities.

The expulsions of left-wing councillors in Liverpool which the Labour leadership had implemented in the 1980s failed to prevent the regroupment of the Militant group and others around Liverpool's Broad Left coalition which campaigned against the moderate party-loyal council leadership. In 1991, the conflict between the majority party loyalists and the Broad Left came to a head when local authority workers went on strike against the council to defend jobs and council services against threatened cuts. The Broad Left fielded its own 'Real Labour' candidate at the Walton Parliamentary by-election on a platform which stressed support for public service employees and opposition to expulsions from the Labour party (Real Labour came a bad third in the election behind the Liberal Democrats and the victorious Labour party). The Broad Left was also against central government incursions on the powers of local authorities and in favour of public expenditure increases to create jobs and promote Liverpool's flagging economy. For the Broad Left, privatisation and the enterprise culture has failed cities like Liverpool.

BRITISH URBAN POLICY

In the early 1990s, therefore, there remained a problem for Labour in that it retained a commitment to job creation and social safety nets, but not at the expense of developing a viable market economy. The discussion of such issues in connection with the political debate about local government is important because the basic principles followed by the Thatcher and Major governments with respect to policy on local government were essentially the same as those adopted in the government's urban policy. There was a strong element of anti-socialist rhetoric, control on expenditure, encouragement of private initiative and the quest for quality and efficiency. It is also important because by the 1990s the Labour party, as in the case of local government, favoured a market-orientated approach and the creation of an environment in which 'opportunity' would flourish in the 'inner cities'.

After the 1979 general election, the Conservative government sought value for money in public programmes and more clearly defined the scope of 'urban policy' so as to achieve greater financial responsibility and a coherent identification of policy aims and objectives. The cooperation of the private sector was again at the forefront of policy concerns. It was also regarded as important that local authorities should cooperate with central government and the private sector in meeting the needs of communities and stimulating urban renewal. However, the Thatcher government's proactive role was not simply motivated by the desire to achieve efficiency and effectiveness in urban programmes. There was also a political impulse which related to the objective of undermining the new urban left and those Labour controlled local authorities which the government regarded as impeding market-led urban renewal.

The development of an identifiable centrally administered 'urban policy' was thus in contrast to the 'arms' length' approach adopted in the USA by the Reagan administration, although many of the policy objectives were similar. The concentration on private initiative and the running down of the role of government in urban policy constituted elements in policy on both sides of the Atlantic (DoE, 1990; Wakeford, 1990). However, the Thatcher approach depended upon a stronger role for central government in developing and 'pump-priming' initiatives which would have a direct impact at the local level. In some cities, it also involved the assumption of economic development and planning powers through the urban development corporations and the creation of enterprise zones. This required substantial commitments of public expenditure and the expansion of programmes designed to assist the development of local self-help (such as support to non-profit organisations, the economic initiatives under the Urban Programme and improvements to training for the unemployed).

Thatcher's Concern for the 'Inner Cities'

This approach was overseen personally by Prime Minister Thatcher who maintained an interest in the plight of 'the inner cities', seen by her to represent a unique opportunity to show the benefits of the enterprise culture (Cabinet Office, 1990). This was underscored by the launching of the Action for Cities programme in March 1988 following the 1987 general election when Mrs Thatcher announced that the inner cities were to be given high priority as part of the Conservative party's long-term electoral strategy to win them from Labour. Public 'pump-priming' was necessary to give a stimulus to the creation of a viable private sector in the inner cities and encourage communities to develop new economic infrastructures to replace those that had suffered as the result of decline and neglect. Then, according to the Tory strategy, the distressed areas could once and for all relegate socialism to an irrelevance.

Like the government, Labour councils (including left-wing authorities) also had an interest in the fate of the inner cities. They were keen to stimulate economic growth through urban renewal and were aware of the electoral implications of success in creating jobs. The regeneration of the distressed city districts presided over by Labour councils was regarded as a priority objective. The national Labour party favoured an approach which would encourage the private sector as well as the development of partnerships between companies and local authorities. While there was nothing new in this, it was contentious for the urban left.

Urban policy was an area in which the tension between the urban left and the government was defined in terms of a battle by councils for resources and attempts to reduce their marginalisation which was the result of the Conservative government's political 'campaign' to regenerate and 'win' the cities. However, the administrative setting was one in which there was a fragmentation of the making and administration of urban policy between contending interests, public authorities and central government departments. This fragmentation enabled skilful Labour councils to 'play the system' and direct their attentions towards a number of strategically important departments or sympathetic junior ministers and officials responsible for making decisions about fund allocations and the development of new initiatives. To the extent that councils benefited from this, there could be said to have been a network in place which allowed even left-orientated councils (such as Labour's inner London boroughs) as well as more centrist councils to access 'the system' to their advantage (North Staffordshire Chamber of Commerce, 1985). Councils developed economic development units which were aware of the complexities of the Whitehall bureaucracy (Mills and Young, 1985).

Administrative Fragmentation

In 1989, the Audit Commission published its report, *Urban Regeneration and Economic Development* which provided a detailed review of the state and workings of urban policy in the UK. The report was mainly concerned with the interface between central and local government and ways in which friction between different agencies could be reduced. The report was critical of the administrative fragmentation in the making of urban policy and drew attention to the conflicts that characterised the relationship between councils and central government. Many local authorities believed that their role was undervalued by central government and saw themselves as increasingly marginalised. Government programmes were seen by councils as a 'patchwork quilt of complexity and idiosyncrasy' which baffled local authorities and business alike. The rules of the game were regarded as over-complex and sometimes capricious, and there was serious organisational compartmentalisation which worked against the achievement of coherence. Moreover, key organisational structures had 'fallen into disrepair' as some partnership schemes failed to operate in practice (Audit Commission, 1989, p.1)

There was therefore a need for a greater consistency of approach in order to overcome these problems. This required a willingness to recognise the appropriate role of local authorities and to grant to local authorities the ability to respond to local circumstances free of unreasonable constraints. The report urged local authorities to organise their efforts in urban regeneration more effectively and to integrate policy choices on urban issues into their main programmes so that councils could act as leaders and catalysts for change. Local authorities could develop their role in education and training; promote confidence in the cities by encouraging leadership and enterprise; improve the local environment and encourage better use of land resources; link disadvantaged social groups to the private sector; and stimulate business activity.

The Urban Group

The Audit Commission report was concerned with the operational aspects of urban policy and the outcomes that had been achieved during the Thatcher years. The report was critical of the attitude adopted by the government in its relations with local authorities. There was only 'the most cursory mention of local government in the Action for Cities documentation' and government initiatives such as City Grant, and the urban development corporations marginalised local councils despite more recent government actions which suggested a more positive disposition (as in *Progress on Cities* (DoE, 1989) launched in March 1989 which emphasised the role of local authorities in partnerships).

There had been an attempt to coordinate urban policy more effectively. The Chancellor of the Duchy of Lancaster, in the Cabinet as a Department of Trade and Industry Minister, had a coordinators role as Inner Cities Minister up to July 1989. After that date, the coordination role was transferred to a non-Cabinet minister in the Department of the Environment who was at the same time Minister for Local Government. The situation was not satisfactory particularly when it was considered that the Welsh and Scottish offices took the major responsibility for urban affairs in their respective jurisdictions. However, in Scotland, there were five main departments of the Scottish Office, the functions of each being more tightly coordinated than in England. Urban renewal and the Urban Programme were transferred to the Industry Department from the housing division of the Scottish Development Department in 1987. The fragmentation of responsibility across several departments in England was a factor

145

which reduced the coherence in central government policy initiatives (Audit Commission, 1989, p.17).

It was also evident from the government's urban policy that there had been a change in the way in which different policy areas were handled and assessed by the DoE. The gathering of programmes in an 'urban group' (or 'urban block') enabled the Department to set and evaluate policy and programme aims and objectives and monitor expenditure in related areas. This gave a more coherent feel to the way in which the department was thinking about its programmes in general and with reference to specific priorities. It also enabled the DoE to provide a clearer evaluation of government spending on related urban initiatives (see Table 6.4).

Of the agencies and programmes, the 1989 Audit Commission report and the DoE *Annual Report* for 1991 concentrate on the following major initiatives:

1. The Urban Programme

The Urban Programme (UP) grant, through the local authorities, aids a variety of projects within the fifty-seven urban areas of special need identified by the DoE. These are areas which exhibit severe problems of urban deprivation (see Chapter 4 for a discussion of measures of deprivation in the UK). Projects funded under the UP are expected to reflect the central UP aims which refer to reviving local economies and social infrastructures, fostering enterprise and improving the environment to improve confidence and investment.

The DoE *Annual Report* for 1991 showed that the UP's initiatives concerning the environment involve schemes such as landscaping, tree planting, improvements to vacant land and open spaces and environmental works on former industrial and commercial buildings. These measures are seen as part of a package of improvements which involve the UP in interrelated economic, social and housing initiatives. Local authorities are invited to submit programmes under strategic objectives designed to address local needs. The balance of expenditure has shifted over recent years to reflect the UP's concern with economic projects and there has also been greater emphasis placed upon capital as opposed to revenue support. This has been within a context that has stressed value for money and the need for project administrators to monitor and evaluate the outcomes of their activities.

2. City Grant and the Derelict Land Programme

City Grant was introduced in May 1988 to replace the Urban Development Grant (UDG), Urban Regeneration Grant and private sector Derelict Land

Table 6.4 Expenditure on inner cities, UK, £ million

	1985–86 out-turn	1986–87 out-turn	1987–88 out-turn	1988–89 out-turn	1989–90 out-turn	1990–91 estimated out-turn	1991–92 plans	1992–93 plans	1993–94 plans
Urban Programme	253.6	236.7	245.6	229.0	226.0	232.6	243.2	243.6	245.9
City Grants	23.7	23.9	26.8	27.8	39.1	49.0	66.0	76.0	78.7
Derelict Land	73.0	78.1	76.5	67.8	54.0	64.3	75.5	83.1	86.8
UDCs (EFLs)	85.6	89.3	133.5	234.4	436.0	550.2	473.1	305.4	302.8
CATs	–	–	–	–	4.0	7.8	8.1	8.4	8.6
Research	0.5	0.7	0.4	0.4	0.6	0.7	0.8	0.8	0.8
Total	436.4	428.7	482.8	559.4	759.7	904.6	866.7	717.3	723.6
of which:									
Central government's own expenditure (voted in estimates)	90.1	95.3	142.7	252.2	464.3	602.4	551.2	397.5	399.0
Central government's support to LAs									
– Grant (voted in estimates)	289.2	271.9	282.4	248.3	238.0	243.9	265.5	270.8	275.6
– Credit approvals	57.1	61.4	57.7	58.9	57.3	58.3	50.1	49.0	49.0

Source: The Department of the Environment Annual Report 1991: The Government's Expenditure Plans 1991–2 – 1993–4, Cmnd 1508, (1991), Section A6, 'Inner cities', reproduced with the permission of the Controller of HMSO
Note: Figures reflect DOE rounding, USA

Grant (within the fifty-seven priority areas). City Grant is paid directly to private sector developers to support capital projects which will benefit areas which have suffered as a result of physical decline and lack of investment. The grant is intended to cover the gap between development costs and values on completion which would otherwise deter developers from initiating projects.

The DoE claims that the grant has encouraged development in the inner cities by levering additional private sector investment into areas of acute deprivation. The grant has assisted projects such as the development of new industrial units, the refurbishment of offices and workspaces and capital support in the leisure and retail sectors. Under the latest Derelict Land Programme, the government has promoted the reclamation of land in England through the Derelict Land Grant (DLG). This is available to local authorities, public bodies, private firms and voluntary organisations. The aim is to achieve the economic use of the land and encourage industrial, retail and commercial activities on reclaimed sites.

3. Urban Development Corporations

The ten Urban Development Corporations (UDCs) in England are bodies established by the government with the objective of regenerating their designated areas. The UDCs are described as 'short-life' bodies expected to complete their work within a five- to fifteen-year period. The UDCs, in 1991, had responsibility for regeneration of around 15,800 hectares of run-down inner city land.

The UDCs were controversial in that they were established with powers over planning and economic development within the jurisdictions of existing local authorities in major towns and cities. The local authorities were, in effect, relieved of their responsibilities for these matters within the UDC designated areas. The UDCs were charged with:

a) reclaiming and assembling sites for development;
b) assisting with the provision of transport and other infrastructure;
c) improving the environment and helping to provide improved social facilities; and
d) offering financial assistance to private sector development.

All this was to be achieved with the maximum possible amount of private sector investment. The government was aware that UDC designation was often seen as a challenge to Labour controlled local authorities which were keen to maintain strong influence over economic development in their cities (Heseltine, 1987). Sheffield City Council, for example, prior to the designation of a UDC in the city, claimed that it was already doing all that it could to attract private sector investment. The Militant-influenced council in Liverpool maintained a suspicious attitude towards the Liverpool Development Corporation and Labour councillors in London took a similarly cautious view towards the London Docklands Development Corporation. It was only in the later 1980s that Labour councils began to modify their attitudes to the UDCs by adopting a more conciliatory view and by making greater use of opportunities for councillors to be represented on UDC committees.

4. City Action Teams and Task Forces

The City Action Teams (CATs) were established to coordinate and target government programmes in 'areas of greatest need' as identified by the DoE. The CATs indicated an official recognition that there was a need for the kind of coordination in urban policy to which the Audit Commission had drawn attention. However, many local officials regarded the eight CATs as under-funded and lacking in authority as contact points between government departments and other public and private sector organisations. Each CAT, with its 'sponsor minister' from departments concerned with urban renewal, is expected to draw up an action plan with the objective of stimulating cooperation between the private sector, local authorities and voluntary organisations. The main concerns of the CATs are with unemployment, environmental improvement and the need to promote enterprise. They each have a small budget which is intended to 'prime' initiatives related to these areas. The task forces were similarly encouraged by the government to develop initiatives to generate jobs and improve communities.

EUROPEAN INITIATIVES

While Labour local authorities felt that the government's urban initiatives were discriminating against them, they were keen to enjoy funds available from the European Community (EC). Local authorities could back effective bids by taking the initiative in impressing upon the EC that they suffered from economic problems which required additional funds over and above those which were available in the UK (Audit Commission, 1991). Labour councils invariably used the EC bidding process as an opportunity to point to what they claimed were the inadequacies of government support for community and economic development initiatives.

The European Regional Development Fund (ERDF) was established in 1975 to help reduce regional imbalances in the EC and was worth 4.5 billion ECUs in 1989 (an ECU being the monetary unit of the EC budget which was equivalent to £0.727 or US$1.202 in January 1990). The European Social Fund (ESF) was worth 3.5 billion ECUs in 1989 and had the aim of promoting workers' job opportunities in EC member states. The European Agricultural Guidance and Guarantee Fund (EAGGF) Guidance

Section (part of the reform of the Common Agricultural Policy) aimed to speed up the adaptation of agricultural structures and contribute to the development of rural areas (worth 1.5 billion ECUs in 1989). The Community also offered loans to support its structural polices via the European Investment Bank (EIB), the New Community Instrument (NCI), the European Coal and Steel Community (ECSC) and the European Atomic Energy Community (Euratom). The aim was thus to focus Community structural action on those regions and areas experiencing the greatest difficulties and on priority fields identified by the Community (Reid, 1990).

EC programmes stressed the need for 'value for money' and the decentralisation of the management of Community assistance. The wording of EC project descriptions often sounded as though it had been drafted by Conservative party central office. There was also emphasis upon the need for recipients of EC funds to monitor and evaluate programme outcomes. Local authorities in the UK were thus obliged to take account of the specifications and objectives of EC programmes. They were also encouraged by EC officials to develop partnerships and support a variety of initiatives intended to stimulate small- and medium-sized enterprises (SMEs) and the private sector in general (Commission of the European Communities, 1991).

CONCLUSIONS

By mid-1991, the Labour party had adopted a set of policies on local government and urban renewal which were less ideological and less socialist than those advocated in the 1983 general election. The degree to which they were 'socialist' in the traditional sense of the term was questioned even by some Conservatives. Labour's 1991 policy statements emphasised quality in the provision of local services and the quest for opportunity in education and social provision. Labour also sought to match the Conservatives with its own version of a citizen's charter. How all this differed from John Major's 'opportunity society' and the Conservative's view of citizens' rights remained a matter for party political debate. The social market was difficult to distinguish from the 'socialist market' advocated by those in the Labour party who still retained a certain collectivist instinct. The Conservative party's alternative to the poll tax, the Council Tax, bore a resemblance to Labour's 'fair rates' proposals despite attempts by both Labour and the Conservatives to draw a distinction between them. In 'new times', old-style municipal socialism had therefore effectively been removed from the mainstream political agenda.

7

INCORPORATING COMMUNITY POLITICS

Chapters 3 and 4 examined the economic factors influencing change in the cities. Community groups operate against the backdrop of economic change. Their attitudes and strategies are influenced by the resources available, the 'ghetto' syndrome often associated with distressed districts and the move of business to the suburban areas. Local interest groups or voluntary organisations have formed to defend the interests of citizens and extend social, economic and political opportunities. In the era of the 'enterprise culture', communities have been faced with new choices about how they can achieve their goals. Ethnic minority organisations have faced special problems. Blacks, and other minority groups, have not only been weakened by their diversity and lack of political cohesion, but have also been affected by a lack of political influence reinforced by racism (Engstrom and McDonald, 1981, 1982; Jacobs, 1986, 1988; Messina, 1987).

PUBLIC ATTITUDES

The motivations of community groups are influenced by a range of factors relating to the political conditions in society as a whole. Public attitudes during the 1980s confirmed the view that during the Reagan and Thatcher years there was a desire to advance by applying practical and achievable policies. In Britain and the USA, there were indications that public opinion was supportive of initiatives which tried to address community problems in positive ways and which mobilised both public and private resources in so doing.

A survey published in 1986 by Social and Community Planning Research, provided an important indication of the attitudes of the public in Britain and the USA to major social and economic issues. The study was produced by James Davis (1986), Professor of Sociology at Harvard University and Principal Investigator for the National Opinion Research Center's General Social Survey (GSS). The study sought to identify any broad similarities in attitudes in the two countries. It included clusters of questions relating to attitudes about: a) police powers; b) government intervention in the family; c) civil liberties; d) school priorities (how people viewed the importance of certain topics); e) social and economic welfare; f) government economic policies; g) government spending priorities; h) political impotence (does the public enjoy political influence?); i) social inequality; and j) gender inequality.

Responses were sought on the basis of 'commonsense' interpretations related to each of the above items. A five-point scale was devised which recorded categorised responses of strongly in favour, in favour, neither in favour nor against, against, and strongly against. The categories were then reduced to 'in favour' (including strongly so) versus all other answers. The same approach was applied in both Britain and the USA, allowing for broad comparisons by country to be made. The basic measure of similarity was thus a straightforward one, measuring how close the two sets of percentages were for each of the items.

Davis found that for clusters a–d, the deviations were small. In items relating to police powers, government intervention in the family, civil liberties and school priorities, there was a marked similarity between US and British opinion. For clusters e–h, the average deviations were greater, with the British being more supportive of governmental provision in the social welfare and economic fields. The British tended to favour public intervention in general, but were shown to lack a sense of identity with government. Both British and American respondents showed similarities when asked to rank their priorities in social policy with provision. Health care, for example, was seen as an area worthy of government support.

With respect to civil liberties and police powers, respondents were asked for their approval or disapproval of four police activities. These were surveillance, telephone tapping, opening mail and overnight detention. In Britain and the USA, there was a marked opposition to opening the mail of criminals, but support for detentions and surveillance (around one-third not favouring detentions in each country). The survey also found that in both countries clear majorities (around two-thirds) endorsed the 'traditional, well-behaved, mild-mannered forms of protest' (Davis, J., 1986, p.98). However, in both nations a fifth to two-thirds of respondents felt that even these forms of protest should be made illegal. This suggested a latent conservatism within the two countries on issues relating to social protest and civil liberties.

Davis found that four clusters fitted a pattern which he described as 'constant differences'. These were cluster e – social and economic welfare; cluster f – government economic policies; cluster g – government spending priorities and cluster h relating to political impotence. Responses showed that the British were more favourably disposed to government responsibility in social and economic welfare provision, with 57 per cent believing that government 'definitely should be' involved in Britain, compared with only 24 per cent in the USA. The most popular programme among Americans (which was assistance to the aged) received around the same level of support as the least popular programme among the British (provision of jobs), each with about 40 per cent.

The popularity of government intervention in Britain did not appear to imply that the British were particularly strong supporters of state ownership. Only 25 per cent of Britons supported state ownership of electricity, but this did compare with only 6 per cent in the USA. For Davis, the conclusion was that the British appeared to favour a stronger welfare state than the Americans, but they were discriminating about the precise nature of state involvement in regulating and owning industries.

Davis was also concerned with social class. National differences were small on the perceived 'amount of inequality' and large on reducing class differences (the British were more inclined to want to level class differences). Twenty-one per cent more Britons than Americans agreed with the assertion that achievement depended upon family background. However, the data failed to reveal any consistent national differences apart from those relating to 'levelling' or any desire by the publics surveyed to adopt radical means to bring about change.

The Americans were more suspicious of 'big business' than the British, but also more hostile towards the trade unions. Considering the coolness of most Americans towards the unions, the responses suggested that, while most favoured individual effort and free enterprise, they were keen to control the activities of the larger corporations in the interests of equity and fairness.

THE END OF CLASS POLITICS?

There were some obvious contradictions in the attitudes of those surveyed such as the American suspicion of 'big business' in a strongly entrepreneurial culture. However, there was little indication that the findings helped to define anything but a relatively low level of class consciousness. There was a marked tolerance on class issues which bore little resemblance to traditional Marxist notions of class confrontation or class warfare. Did these findings therefore confirm the breaking down of class politics?

A New Political Culture?

The idea of a 'new political culture' (NPC) was developed to highlight the changes in political attitudes and the conditions which have affected them. Clark and Inglehart (1989) argue that the emergent NPC builds upon the new policy preferences of citizens and political leaders. It encompasses aspects of political expression commonly associated with green, yuppie, populist and other politics. The NPC has challenged class politics despite the maintenance of class as one important element in the

NPC. Clark's framework incorporates a limited version of class politics in which the rich and poor continue to express conflict through political parties. For Clark and Inglehart, the core of traditional class politics was that it was based upon the assumption of a hierarchical division of society and that class cleavages explained political cleavages and voting behaviour. A major issue in class politics was that of redistribution and the degree to which governments acted effectively and fairly in handling the distribution of income.

Clark and Inglehart refer to the concept of 'the new class' which emerged during the 1970s. This recognised the growth of a new young and well-educated group of middle-class voters which was substantially more liberal than most other voters. Economic and social changes were responsible for changes in voting patterns and the attitudes of the electorate as a whole. Changes in employment patterns and population shifts were also contributory factors in producing new political motivations and alignments.

In this situation, class politics became inadequate when faced with new patterns of social stratification. Class never explained 'all the variance' (Clark and Inglehart, 1989, p.5), yet it was often regarded as the best 'tool in the tool box' for viewing such stratifications as class-related phenomena. 'New' cleavages emerged around issues such as gender, race, regional loyalty, sexual preference, ecological concern and so on (Clark and Inglehart, 1989, p.5). Clark and Inglehart argue that the core of these issues was not economic or redistributive, but essentially social. Consequently, the NPC displays the following characteristics:

1 *Social and fiscal–economic issues are explicitly distinguished.* Social issues demand analysis in their own terms. Social issues as perceived by community leaders and politicians can no longer be derived from their positions on fiscal issues. Leaders and parties can no longer rely upon economic issues to win them support.
2 *Social issues have risen in salience* relative to fiscal–economic issues.
3 *The classic left–right dimension in politics has declined.* New combinations of social and fiscal policy preferences have emerged among citizens and political leaders which make 'post-industrial' society more difficult to manage.
4 *Market individualism and social individualism have increased.* This has generated problems for the left and has challenged reliance on welfarism and public ownership.
5 *The welfare state has thus been questioned.* New issues about service delivery and the efficiency of public provisions have emerged, together with policies which have promoted private tendering and the notion of the citizen as consumer.
6 *The rise of issue politics and the decline of hierarchical organisations.*

Traditional bureaucracies are under pressure as new social movements have developed around social issues. Traditional hierarchical parties, trade unions and government agencies are increasingly seen as outdated and inefficient and even undemocratic. 'Intelligent citizens' refuse to be treated as the docile subjects of party officials or government bureaucrats.

7 *These NPC views are more pervasive among younger, more educated and affluent individuals and societies.* The NPC has developed along with basic changes in the economy and the family. The politics of the NPC has encouraged a broader value consensus which gets away from the old politics of social and economic hierarchy.

GROUPS AND THE NEW POLITICS

This is very persuasive. There is much in the NCP analysis which accurately describes community-level politics in the present period. However, there are a number of qualifications which need to be made. First, while it may be the case that 'class' politics is on the decline in the sense that there is less direct public identity with traditional class labels, there is little that is 'new' about this in terms of political analysis (as Clark and Inglehart admit). The idea that the working class was abandoning class affiliated politics was one that was common in the political science literature of the 1950s and 1960s (Kelley *et al.*, 1985). Studies of voting behaviour in Britain tended to support the view that class was a significant factor in determining the patterns of allegiance to the Labour and Conservative parties as late as the 1970s and beyond, but there was an increasing interest shown in the 1970s in the reduction of the importance of class-defined policy issues. In the USA, research pointed to non-class voting on racial issues (Sheffield and Hadley, 1984). Theories of 'embourgeoisement' claimed that the working class was becoming more like the middle class in the new age of consumerism and was thus more likely to break with a class politics which stood for the defence of the interests of a socially specific section of society (Hindess, 1971).

However, even accepting such social changes, the political outcome could also be viewed as the consequence of *a redefinition* of class politics as opposed to its *end*. In Britain there was evidence in the 1990s that class identifications and even collectivist attitudes could re-emerge (Crewe, 1990). Inglehart (1985) concedes that the orientations of western publics can show 'impressive durability'. Wildavsky (1987) argues that certain basic preferences (such as favouring egalitarianism) may endure even though beliefs about how to achieve them may change. Left–right interpretations of public attitudes can be misleading if they are misread as indicators of deeper changes in the political culture. In this context, working-class political expression may better be seen as having been recast as it

155

became incorporated into 'mainstream' politics and the 'legitimate' policy-making channels of democratic capitalism and as political conditions changed (see Chapter 1).

A second cautionary note concerns the question of how distinct *social and economic issues* may be analysed in a changing political environment. It is very difficult to make a hard distinction between social and economic issues or to identify specifically 'fiscal' issues inside broader 'economic' ones. The NPC position does not deny the importance of broader economic issues. However it poses the critical question that given that social concerns have become more important in public opinion, how, and to what extent, do economic–fiscal questions now define the parameters of major debates about them?

As indicated in the NPC position, the debate concerns the place of 'the economic' in political life. On this issue, the cautionary note relates to the need to view 'the economic' as something that is complex and pervasive when viewed against the change in its salience. Public perceptions may stress social concerns, but in practical terms these are difficult to separate from economic factors. For example, the search of American and British charitable organisations for resources has been one that has involved *the meshing of social and economic demands* by groups (Ware, 1989). McCarthy's study of the Child Poverty Action Group (CPAG) in Britain stresses the economic questions facing a major interest group campaigning on behalf of the poor in spite of the distinction made between social and economic issues in official public policy (McCarthy, 1986, p.3). The CPAG expanded in response to the 1964–70 Labour government's inaction on social issues when there was an expectation that the Labour party would introduce policies that would benefit working-class households. It was the CPAG's campaigns which sought to combine a social concern with a policy which called for increased expenditure on social provisions as the only effective way of initiating an effective anti-poverty strategy at the national and local levels.

More recent conflicts between groups and the Thatcher and Major governments over reforms in the National Health Service, education and the poll tax also involved strong links between social and fiscal/economic concerns. The attempt of interest groups to resist reductions in health service provisions were associated with the government's desire to introduce market forces into health provision. There was a similar policy of marketisation where schools were encouraged to become more aware of their resource bases and operate in line with clearer economic objectives.

Issues of major public concern involve decisions about resources, and this involves groups in competitive policy settings where the struggle for economic resources are *ultimately* the basis of the political agenda. This even exists where there is an 'apparent' public policy distinction between social and fiscal issues (McCarthy, 1986, p.3).

City Trenches

An element of *conflict* in community politics is implicit in the competition for resources between different groups which tends to undermine 'harmony' and stability. By placing more emphasis on conflict, a more complete view of the change in class politics may be possible.

Katznelson (1981) is concerned with the recasting of class politics under different political conditions. For Katznelson, it is a question of the changing nature of class politics rather than about 'the end' of class politics. Class is seen to be the basic point of reference in an analysis of community politics and one which helps to define the different forms of class-based political activity. In this frame, an absence or reduction in a specifically working-class consciousness reflects the manifestation of a new form of politics characterised in terms of a non-collectivist ideology potentially detrimental to working-class interests.

Katznelson views class politics in its historical perspective and compares the USA with Europe:

> Industrialisation and working class history throughout the west have been mainly urban phenomena. Workers have experienced industrial capitalism as dwellers in the cities where they have laboured and where they have lived. To a very large degree differences in the national histories of working classes reflect the different ways in which workers have understood the relationship between urban workplaces and residential communities.
>
> (Katznelson, 1981, p.7)

Katznelson acknowledges 'American exceptionalist' theories which stress the peculiar features of American working-class politics which developed a less collectivist orientation than that adopted by European labour movements. America's working class was not created 'at once', but was fashioned in terms of the national, ethnic and religious affiliations of workers. The composition of the American working class was determined by three movements of population – from country to city, from east to west and south to north, and from Europe and Asia and the Caribbean: 'The result has been an incredibly polyglot working class, differentiated culturally, racially, economically and spatially' (Katznelson, 1981, p.10). This made class solidarity difficult especially during times of economic restructuring and spatial reorganisation (a process which continued during the era of high technology and urban expansion). Racial antagonisms added to the difficulties of engendering a collective working-class consciousness. However, this provides only a partial analysis.

Ethnic and racial ties sometimes stimulated collective class activity while rapid industrialisation in the USA produced class-forming coalitions. Katznelson suggests that the analytical marriage between urbanism and American exceptionalism 'may produce the progeny' needed to provide a clear statement about the nature of American class politics. The main elements of explanation for the boundaries and rules of American urban politics and for 'what has been special about class relations in the United States may be found in the ways in which workers understood the objective separation of work and home in early industrial cities and acted on that understanding' (Katznelson, 1981, p.17). Therefore:

> The objective separation of work and home in the antebellum city presented American workers with three logical empirical choices. They might have come to see themselves (as workers did in England, for example) as workers not only at work but also at home; or (as in the 'plural societies' of Belgium and Holland) as ethnics at work and ethnics at home. Although these two configurations were in fact lively possibilities at the time, they gave way to a third, and distinctively American pattern. The two main European definitions of work and home used the same co-ordinates for both realms. In the United States, by contrast, they were quite separate: most members of the working class thought of themselves as workers at work but as ethnics (and residents of this or that residential community) at home.
>
> (Katznelson, 1981, p.18)

The implications of this are examined with reference to class solidarity. There is a solidarity at work, but there are also ethnic and territorial frames of reference. The politics of work and community have their own vocabularies which take on sharply divided forms in the USA. This system of values and 'customary practices' provides 'the main political formula of ideas, organisations and activities that has protected the core arrangements of capitalism in the US from challenge' (Katznelson, 1981, p.19). Drawing on the work of Antonio Gramsci, Katznelson argues that in the advanced industrial societies such clusters of ideas and behaviours are 'like the trench systems of modern warfare' where the trenches define the terrain of battle between groups. Each of the 'trenches' is distinctive in defining the place and content of conflict.

The trenches are defined by their country-specific social and political system characteristics which delineate what is special about class and politics in each society and which help to shape the 'rules of conflict':

> In the United States the most important set of rules has been urban. This American urban-class system of 'city trenches' has defined what is exceptional about class (and race) in the United States, and it has made very difficult the emergence of socialist, social democratic, or labour parties on the European model of the late nineteenth and early twentieth centuries.
>
> (Katznelson, 1981, p.19)

Since this was written, accelerated change, particularly in Europe, has focused attention on the breaking up of the traditional socialist orientation. This has been manifest in the decline of trade union membership, the atomisation of local politics and the emergence of the politics of the 'new times'. Some of the class perceptions associated with the USA are now more prevalent in Britain.

The Katznelson model interprets changes by asserting that working-class politics has undergone a political shift in favour of capitalism. Workers' subjective perceptions of class are said to be out of line with their objective interests. The analysis concentrates upon the *motivations of social classes in political flux*. The focus is on classes in relation to each other within the economic and social order of capitalism. The substructure of this order is concerned with relations between classes and their ability or otherwise to assert their authority over economic resources. This is a conception which sees changes in political conditions taking place inside the broader framework of capitalist economic relations which essentially remain intact in spite of modifications to the societal superstructure (which includes the attitudes of classes, parties and so on).

CLASS, 'COMMUNITY' AND POWER

There are important implications in this for the ways in which 'community politics' and the role of social class are viewed. The Katznelson approach disrupts the conception of 'community' as one that involves a harmonious interplay between different interests. Katznelson refers to Raymond Williams who argued that the 'community' never seems to be used in an unfavourable way (Katznelson, 1981, p.26; Williams, 1976). The connotation is generally one that implies that 'community' is something positive and something which counters the negative aspects of capitalism.

This romantic view of 'community' is one that is seriously flawed. Even if 'community' were to be defined with reference to a spatial entity, it would be possible to point to developments within community 'jurisdictions'. The evidence consists of references to riots in the 'inner cities', the rising crime rates in urban localities, conflicts between racial and ethnic minorities and whites and the prevalence of class divisions. 'Community' organisations reflect the divisions within society and even where there have been attempts to create a united 'community' voice there are tensions which threaten to destabilise coalitions and lead to renewed political conflicts (Jacobs, 1988). 'Community politics' (if this is to be used as a convenient coverall term) therefore involves conflict between classes and between citizens and elite groups.

POWER AND ACCESS

Having acknowledged the transformation and 'incorporation' of class politics in general, there still remains the problem of how analysis can account for the practical incorporation of community leaders into the policy process. Those who see local-level politics as being about the interplay and reconciliation of conflicting interests have pointed to the importance of elections in making political leaders accountable to local electorates (Dahl, 1956; Newton, 1976) and encouraging cooperation. However, as Newton states, there is room for doubt whether local elections do actually make leaders pay attention to the wishes of the public. There is no direct transmission mechanism which connects public preferences to final policy outcomes and local policy-makers enjoy a considerable degree of autonomy when it comes to policy formulation and implementation (Dearlove, 1973; Newton, 1976).

Almond and Verba in their classic 1963 study, *The Civic Culture*, understood that there was an important arena of politics outside the legitimate electoral sphere. They saw voluntary organisations as the most important means by which individual citizens could come into contact with local government officials and local politicians. This was in a pluralist setting in which group activity was an important way of expressing policy preferences rather than through direct electoral activity.

The Role and Problems of Leadership

The pluralist view, however, was inadequate. It failed adequately to account for the role of political leadership in accommodating within legitimate political structures. Policy-makers favour a stable political environment so that they may formulate and implement policies effectively and in a democratic manner (Miller, 1983). The corporate private sector also likes stability which is conducive to the successful pursuit of commercial and industrial activities. In a democratic society, governments are therefore aware of the need to encourage interest groups to participate in a variety of ways within the political system and cooperate with public and private agencies where appropriate.

In even the most distressed urban districts, there are many political and economic incentives for groups to cooperate. These may arise from government initiatives, public–private sector partnerships, government grants and so on. The leaders of interest groups may find such incentives attractive and decide to apply for funds or work with public and private sector bodies to promote community or group interests.

Most of the time there are few manifest problems with such arrangements. Generally, moderate groups are happy to operate in this kind of environment (Newton, 1976). However, local leaders may sacrifice their

160

independence by entering into relationships with government. They may also tend to drift apart from the 'constituents' which they claim to represent, particularly as they become professionalised and incorporated into legitimate political structures. This does not necessarily imply that activist leaders are always more radical than the general public (Welch and Studlar, 1983), but it does suggest that public disaffection may arise out of unrealised promises, high expectations or fundamental shifts in the public mood (Duckworth, 1983).

Leaders who get elected to office at the national and the local level often find that the realities of office impose constraints. Such problems constitute part of the crisis of political leadership where the aspirations of 'the people' can often come seriously to diverge from those who represent them. More specifically, there are several elements which together make up the leadership problem in distressed urban districts in Britain and the USA:

1 The incorporation and professionalisation of community representatives into various formal and informal structures and/or their election to political office during a period in which resources are limited and government expenditures constrained. There are always risk factors in political coalitions under such circumstances (Wright and Goldberg, 1985).

2 The tendency for the realities of office or participation (and the incrementalism of policy-making) to blunt the impact of well-intentioned politicians (Dearlove, 1973; Jacobs, 1986, 1988) thereby leading to the dissatisfaction of supporters (particularly where ideological or long-held commitments have to be compromised or abandoned).

3 The tendency for community leaders to modify their demands to placate government officials and politicians or interests 'outside their electorate' (Carson and Oppenheimer, 1984) – such as trade unions or corporate interests – in order to gain political favours or retain access to the political system.

4 The manifestation of social disorders in communities suffering from economic disadvantage and racial tensions (see Chapter 8) where traditional political arrangements are placed under pressure by the actions of 'the unorganised' on the streets (see Muller and Opp, 1986 who see riots as economically motivated) or where, in the most extreme cases, the authority of community leaders and the police is challenged by new political movements or possibly by organised crime such as drugs rings (Henry, 1987, pp.508–28).

THE INCORPORATION OF COMMUNITY POLITICS

A key factor is the extent to which 'community politics' has been incorporated into legitimate political channels thus requiring moderate leaders to make compromises with established politicians and public officials so that they may increase their chances of obtaining scarce resources. Incorporation may involve one or more of the following:

1 the attainment of local government representation;
2 various degrees of participation in government and other public agencies;
3 involvement in public–private partnerships and/or other forms of formal and informal access to the political system.

In Britain and the USA, the influence which the 'moderate' leaders continued to have depended upon the maintenance of their often partial, sometimes comprehensive, accommodation within legitimate political structures. Local organisations of all kinds tended to recognise the advantages of working within the system. Many came to be directly or indirectly involved in policy formulation and implementation. This enabled community organisations to obtain tangible, even though sometimes limited benefits for their 'constituents'. The benefits of incorporation were important in encouraging groups to cooperate with central and local governments. Even though incorporation did not necessarily involve formal representation at the local level, it did mean that some local organisations were generally better placed to convince their followers of the benefits of compromise and consent in politics; at least enough to ensure the continued existence of good relations between community leaders, politicians and local and national government bureaucrats.

In the 1980s in Britain and the USA this provided the setting within which the political left came to a compromise in their dealings with Conservative governments. As indicated in Chapter 1, the orientation of the left towards the acceptance of private initiatives linked with the notion of the socialist market and the need to realise the role of social individualism. This provided the opportunity for leftist organisations at the local level to justify cooperation with the private sector and develop a range of community-level initiatives which were consistent with the privatisation of urban policy.

Elites and Power

Friedland's (1982) study of *Power and Crisis in the City*, provides a provocative view of the role of elites and interest groups and challenges the pluralist characterisation. Friedland concentrates upon the role and power of business corporations and trade unions. For Friedland, the power of national economic organisations such as corporations and labour unions

162

does not originate in the participation of business or labour leaders in American government. Rather,

> its origins lie in the organisation's control over resources – private investment and an apparatus for mass mobilisation – upon which government depends. The participation of corporate or union elites is more a consequence than a cause of such political power. Political power may be silent, voiceless.
>
> (Friedland, 1982, p.1)

The combination of a changed class politics and the redefinition of the relationship between social and economic issues has produced profound changes in *the forms* of both the American and British political cultures. However, the above considerations indicate that changes should not be viewed in 'black and white' terms. Social and fiscal questions may be blurred and related in complex ways. Community leaders tend to orientate towards the major parties and elite groups which act as reference points for their political demands. Leaders want to increase the economic resources going to their communities and community leaders remain as the links between the locality and city hall bureaucracies (Jacobs, 1986).

There are both benefits and costs involved in this relationship for interest groups, but established party and local government structures are vitally important 'targets' in the complex of local politics. If the local political process is defined in terms of groups seeking resources from governments and the private sector, groups seek entry to the system from without. 'Insider' groups enjoy their status because they have been prepared to sacrifice their total independence in return for tangible benefits (Dearlove, 1973; Olsen, 1976).

The responses of interest groups to policy issues, such as education, health and poverty, thus need to be regarded with reference to the special role of group leaderships which maintain links with parties, public bureaucracies and the private sector. Community leaders do not simply react to changes and events in a way which exactly reflects changes in some wider 'political community'. Instead, they adopt attitudes and political strategies which are influenced by changing social and economic conditions. They are influenced by their *differing interests, class backgrounds and motivations.* As seen above, the 'community' comprises numerous groups which may be in conflict or in coalition depending upon specific local circumstances. In 'new times' the emphasis has been upon coalition, partnership, and incorporation, and in this the leaders of individual interest organisations, amongst other things, may act in a number of ways:

1 As links with local and national political parties and public bureaucracies in bargaining for resources (McCarthy, 1986) and promoting the

demands of their organisations and/or their wider constituency (Whiteley and Winyard, 1984).

2 Taking decisions on behalf of their members or 'constituents' which
affect the strategies of interest groups within the policy process.

3 As intermediaries between their 'constituents' and public and private
agencies involved with particular policy issues.

4 Gaining representation or election to bodies established to handle such
issues or promote community-wide partnerships between the public and
private sectors (Jacobs, 1986, 1988).

Networks and 'Policy Communities'

This view of the role of community leadership is one that places emphasis
on the accommodation of leadership within the political system. This
differs from a pluralist conception of the role of leaders: here groups are
seen as having to make important political compromises in order to attain
access to politicians and officials. Groups often have to work hard at the
interface between the wider political environment and the political system
(Dearlove, 1973). However, the view adopted here is not corporatist either,
despite the concern for incorporation.

It is instead more appropriate to take account of the complexity of the
political system. There is a variety of possible forms of relationship
between interest groups, government and the private sector. Policy networks exist which are a complex or cluster of organisations which link
together public and private interests. Such networks have different structures and display different relationships between groups and governments
(Rhodes, 1986). These differences vary according to the memberships of
the networks, the dependency relationships inherent and the degrees of
integration within the networks.

Rhodes (1986) distinguishes between five types of network which range
between highly integrated 'policy communities' to loosely integrated issue
networks. The policy communities are characterised by the stability of
relationships and the continuity of a restrictive membership. They are
vertically interdependent and are based on the major functional interests
of government such as education, social services and so on ('territorial
communities' encompass interests based on geographical areas such as
Scotland, Wales, etc.). Other networks exist. There are professional networks, intergovernmental networks and producer networks which all share
the characteristic of clustering interests that have a common focus or
policy concern. As Stoker (1988) argues, an appreciation of the political
system is thus possible, not in terms of pluralism or corporatism, but in
terms of a view which stresses the range of possible relationships between
groups and government in different policy areas.

This is especially relevant in a study of urban politics. Chapter 8 indi-

164

cates that black organisations have found it difficult to gain representation at the local level in Britain. In the USA, political representation is wider for blacks. Chapter 9 describes the variety of arrangements which have developed in public–private sector partnerships and the role of business leadership in a number of different contexts. There is thus an 'open' political system to the extent that it is free and democratic. This is a diverse system in which access is limited by certain rules of the game and where policy networks and communities take different forms.

INCENTIVES TO COOPERATE

Governments have managed to extend the range of inducements on offer to group leaderships and thereby have attracted support for urban policy initiatives, but this is not a cost-free exercise for groups. The advantages inherent in a cooperative stance are sometimes offset by significant disadvantages. The costs may be substantial enough to alert groups to the disincentives to cooperate, although the assessments made of potential incentives and disincentives are highly subjective. In practice, this means that the perceived costs of a particular strategy for one group may not register as a cost factor for a different organisation. Group assessments of the perceived costs and benefits of cooperation are thus seen in terms of their own specific 'market' situations vis à vis public and private agencies, and within the context of the particular local political situations which they face.

Incentives and disincentives are of concern to both leaders and followers. Mancur Olson (1965), in line with the public choice conception, sought to explain leadership roles in terms of the importance of the individualised incentives offered by groups to members. He concentrates on collective and selective benefits, with collective benefits being received jointly by group members and followers. Selective benefits are available to the paid-up members of groups. Participation in groups is seen not to be dependent upon the provision of collective benefits through group activity, but upon the more selective benefits arising from individual participation and from the range of sanctions which groups can impose in order to get members to comply. Olson saw this to be applicable to trade unions and professional organisations where group leaders were often pursuing goals which were out of line with the individual interests of the members.

There is value in this view insofar as Olson draws attention to the leadership role and the impact of leadership decisions upon group members and constituents. There is, however, an overemphasis in Olson upon the importance of individualised incentives which leads to difficulties in explaining why it is that large numbers of people involve themselves in collective activities such as strikes, demonstrations and large interest groups (Dunleavy, 1991). Hansen's (1985) research, based on US experience, indicates that collective benefits play a more important part in influencing group participation. Hansen suggests that collective benefits can be significantly expanded by way of public or private subsidisation of groups. It is logical to expect group leaders to maximise both collective and selective benefits for group members, the broader community and themselves.

Both kinds of benefit are implicit in the following examples, and this should serve to underline the inter-play between the two types of benefit. The use of benefits and costs implies that there is a value in a rational choice model which develops an understanding of the complexity of group motivations and the various factors which influence group membership (Dunleavy, 1991).

Economic Incentives

Many economic incentives exist of which community organisations have become increasingly aware, particularly within governments' small business and urban programmes (see Chapter 9). In the USA small business programmes, often directly linked to the major private corporations, have been of importance in regenerating communities and stimulating business enterprise at local level. There is particularly strong involvement with the private sector in the USA, with many black organisations joining programmes which have distinctly non-radical social objectives.

Legitimation

Incorporation into the 'legitimate' policy process can be a powerful political experience for local leaders who may not be used to being regarded positively by local officials and politicians. The status bestowed upon them may be regarded as being of more value to a group than purely economic incentives (Olsen, 1976). Some black organisations see legitimation through political incorporation in terms of having been 'bought out' by the white political structure. With such groups this incentive is, at the very least, of no value.

Some groups actively adopt disruptive tactics to make their point. Recent actions by the US Aids Coalition To Unleash Power (Act Up) exemplify the proliferation of 1990s-style 'direct action' groups. Act Up represents gay rights, pro-abortion and Aids activists who believe that the US government, the 'scientific establishment' and the pharmaceutical industry do not care about the fate of Aids sufferers. The response from Act Up has been public demonstrations and acts of public disruption and violence.

Environmental groups have often found it difficult to gain a good reputation in government either because their demands appear to be impractical or because they have adopted 'unhelpful' attitudes (Dearlove, 1973). Friends of the Earth was formed in the UK in 1971 and was for many years regarded as a 'fringe group' by some politicians. However, by the 1990s the group was an active lobby organisation with sympathetic contacts in many local authorities and public agencies. Its director, Jonathon Porritt, became a television personality and advocate of getting green issues out of the media's 'ghetto slot' (Porritt and Winner, 1989). Other groups, such as the Town and Country Planning Association, have enjoyed a longer record of 'respectability' (McRobie, 1990), long pre-dating the 'greening' of the Thatcher government in the 1980s.

The process of legitimation generally assumes that groups will be treated without discrimination, and so affirmative action policies are important in this context. The experiences of American community action programmes suggests that even radical organisations can be encouraged to participate if they feel that local programmes are conducive to enhancing their legitimate position within the community.

Professionalisation

Participation in government and private sector projects provides groups with an incentive to reorganise their administrative structures and adopt new and more efficient practices (see Chapter 10). Many government programmes are adopting management training as an integral part of community group professional development and non-profit organisations have been active in improving their administrative and managerial practices.

Many groups will see this as a positive development, while others remain suspicious. Professionalisation can be regarded as an objectively positive incentive to groups largely because it enhances resources and enables groups effectively to present their demands to policy-makers. It also enables group leaders to maintain control over resources.

THE COSTS OF COOPERATION

What of the costs of cooperation? Too close a relationship with government and/or the private sector can serve as a potential disincentive to both moderate and militant organisations.

Close Association

The importance attached to association varies, but even where there is a strong commitment to a government- or private sector-backed programme, there often remains an underlying suspicion of the motivations of the supporting agency. As indicated above, some groups see legitimation in negative terms. Groups may not want to sacrifice ideological or strongly held commitments which would be compromised by associating with government or private interests. Marxist organisations have commonly taken this view, together with religious groups and even animal rights advocates.

Some organisations see their objectives being met most effectively by keeping a distance from government, especially where there are political gains to be made. The All Britain Anti-Poll Tax Federation (ABAPT) in 1990 maintained a consistently hostile attitude towards the Thatcher government and was accused even by some on the left as being dominated by extremists intent on following violent and disruptive tactics (Campbell, 1990). Other groups have even regarded violence as a form of 'civil resistance' (Davis, T., 1986).

An identification with physical and police control measures can be an important issue for some organisations. In race-related disputes, accusations of 'racism' abound when the subject of control is discussed, and a distrust of white authority is often evident when the role of the police is under scrutiny (Assefa and Wahrhaftig, 1989). Allegations of police racism appear to be the stock-in-trade of some radical groups, and police tactics are frequently at the core of arguments between local authorities and black groups following riot situations (Jacobs, 1986). This point was illustrated during the 1981 and 1985 riots in Britain, and recently in the USA (see Chapter 8).

Loss of Leadership and Control

Associated with the identification of groups with governments is the problem of loss of leadership control (Dunleavy, 1991). This may involve the leadership being challenged with respect to its general control over group affairs. It is possible that this could occur where a government or other agency lays down burdensome monitoring arrangements over group activities, undermining group independence (see Chapter 10).

THE NEED FOR CONSENT

The extension of incentives is of importance in obtaining the assistance and support of community leaders but it provides no absolute guarantee that the wider community will become involved in public and private programmes. Consequently, many of the newer programmes lay great emphasis upon the broad involvement of the wider community (see Chapter 9). This has the effect of widening the 'catchment' of programmes and implies the need for consent in the development of urban policy. Decentralisation lays heavy emphasis upon the ability of communities to take charge of their own affairs and achieve a degree of economic and political self-confidence.

Local and national government expenditure constraints and over-centralisation have produced a cautious response by groups to issues as programmes in education, social welfare and housing come under pressure. Where economic incentives are limited, local leaders have attempted to promote their interests, not by disrupting local political arrangements (for that could isolate them), but by trying to maintain community resources. Radical political organisations are able to point to the economic consequences of unemployment and inner-city decline. For them it is the lack of incentives which is of prime concern. Their subjective assessment of the costs of political participation thereby concentrates upon the negative impressions of legitimate political integration.

A wide variety of moderate leaders operate with what they regard as a realistic conception of what is possible within the context of the prevailing political system, largely because the wider community itself adopts a pragmatic view of social and economic problems. However, the continued decline of major urban centres may in fact exacerbate the resource problem and help to perpetuate political demands centred around both defensive and radical group political strategies.

8

MINORITIES: 'EMPOWERMENT' AND DISSENT

Ethnic minorities in Britain and the USA have developed an extensive range of organisations to represent their interests. There are social, educational, economic and political organisations which work with national and local governments to promote urban renewal and social improvement. In general, the political articulation of urban ethnic minorities is reflective of a situation in which their communities are located in the most distressed districts where the impact of the 'high-tech political culture' of the 'boom' towns represents something remote. These are districts where local governments have been constrained by budget cuts and falling tax bases. To the extent that urbanisation has impacted on these communities, it has frequently been defined in substantially negative terms, attracting resources and investment away from them to out-of-town locations and the new growth centres in the expanding suburbs (see Chapters 3 and 4).

Ethnic minorities in Britain did not experience the effects of the kinds of upheaval that shook American black communities in the 1960s. There was no British equivalent of the violent urban riots or the mass civil rights movement of that period which led to a major reassessment in the USA of the role of blacks in political life. Britain's assimilationist policies in the 1950s and 1960s were not comparable to the openly racist and exclusionist policies faced by American blacks in the segregationist states of the south. Despite this, there were serious disorders in the UK in the 1970s and 1980s associated with race-related issues and the general condition of blacks and Asians in distressed urban communities.

EMPOWERMENT AND REPRESENTATION

Social tensions and racial discrimination encouraged governments to take account of the problems of minority communities, especially where tensions were likely to produce social conflict and unrest. Ethnic minority organisations were therefore increasingly incorporated into the policy process (Jacobs, 1986). The legitimation of ethnic minority politics also involved the striking up of a series of relationships between groups and the corporate private sector (see also Chapter 9).

Incorporation is often linked with 'empowerment'. This has been particularly so in the USA where blacks have entered into mainstream political life and attained representation at all levels of government. In 1991, the President nominated a conservative black, Clarence Thomas, to sit on the Supreme Court (a decision opposed by many blacks), replacing the liberal black Justice Thurgood Marshall. Blacks have advanced their representation in public and private agencies which deliver services and foster economic growth. The professionalisation of American urban renewal agencies has enabled many local activists to achieve prominent positions within agency managerial structures and local partnerships. To this extent, black politics has been extensively incorporated in the USA and in the UK.

In both the USA and the UK, there exist political rivalries between groups which divert attention away from their common interests. There are special problems in gaining effective representation for ethnic minority views especially where local public bureaucracies are dominated by white officials who are not always sympathetic to minority demands. Controversy thus abounds when race is on the agenda both at the national and the local level. For example, in 1991, President Bush promised to veto a new civil rights bill, backed by the Democrats, which he saw as likely to institutionalise racial quotas in employment. The issue served to illustrate that popular passions about race could still be aroused a quarter of a century after the campaigns of Martin Luther King.

Formal representation is often a factor affecting the extent to which communities obtain support from the public sector for local needs (Jacobs, 1986, 1988). It can also act as a morale boosting factor in circumstances where racism is evident. Democratic political representation is one mark of the degree to which minorities can become legitimated within the political system and take part in local and national government.

The Weight of Minorities

The influence of ethnic minorities in government partly depends upon their numerical influence and the degree to which they are organisationally and politically significant within particular city jurisdictions. Important in this connection is the way in which US blacks are identified as a 'community' which enjoys substantial electoral and political influence as a result of its collectivity and sense of identity.

'Black politics' in the USA implies a conception of the collective notion of a 'black community' which differs from the atomised politics of black, Asian and other ethnic minorities in Britain. In the USA the black population is commonly conceived of in terms of the 'Afro-Caribbean' ethnic-related American community being a separate entity from other black ethnic minorities (such as Asian-originated minorities and even more recent immigrants from Caribbean countries). Between 1970 and 1980, the black community in the USA increased by 17.3 per cent from 22.6 to 26.5 million. In 1980, blacks represented about 12 per cent of the population of the USA. In seven states blacks constituted more than 20 per cent of the population, and there were twelve states in which the black population numbered one million or more. New York had the largest numerical black population (1,784,337) and Washington DC's population was 70 per cent black (Matney and Johnson, 1983).

The 'political weight' of blacks has, therefore, been felt in a very dramatic way in the USA. The American black community has been more politically influential than the geographically dispersed and fragmented black and Asian communities in Britain (Jacobs, 1986). The increased representation of blacks in US national, state and local governments has been an important political development over the last 25 years, following the nationwide campaigns by blacks in the 1960s for civil rights and the end of overt discriminatory practices (Williams, 1987). Without substantial numerical strength, the black civil rights movement would probably have had little impact upon national politics.

Britain's minority communities are generally found in the run-down areas of the larger cities such as London, Birmingham, Liverpool and Manchester. Ethnic minority groups have tended to concentrate in areas where they have been best able to maintain their identities and customs. Individual minority communities and groups tend to be widely distributed around the country in such a way as to reduce their effective strength on an area-by-area basis. This contrasts with the USA where black Americans constitute a more visible presence in society. In addition, in the UK, there is a tendency for small groups of, say, black people, to cluster with small groups of, say, Asians, in one district. This 'atomisation' of black, Asian and other groups tends to reduce their political weight especially in areas where they are unwilling to form multi-ethnic coalitions (Jacobs, 1989a).

In Britain, there exists a variety of organisations representing Britain's 3.5 million non-white population, and its multiplicity of different ethnic communities. In some instances, there is conflict and even hostility between these groupings. Even so, there is a degree of common black identity in Britain which could be the basis for the development of stronger political links between ethnic groups. However, this has not so far been sufficiently strong to consolidate a national identity conducive to the formation of an inclusive black or multi-ethnic political interest group (Fitzgerald, 1984).

Hispanic 'Power'

In the USA, there has been a strengthening of Hispanic (or Latino) identity as the demands for immigrant labour have attracted workers to the USA and Canada. It is estimated that Hispanics could become the largest non-European component of the US population in the twenty-first century, thus giving them a major influence on the future development of American politics and culture. Los Angeles presently ranks as one of the largest Latino cities in the world with 28 per cent or 3.7 million of the population being of Latino origin. In the New York metropolitan area there are over 2.3 million Latinos mainly of Puerto Rican and Dominican origin (Turner, 1990).

Invariably, Hispanics live in some of the worst housing and are employed in low-paid jobs. Since there are many ethnic differences, the term 'Hispanic' covers groups from a variety of national origins. These groups have different ethnic festivals and are unevenly geographically distributed. Cuban-Americans, Puerto Ricans and Mexican-Americans have different backgrounds although many Mexicans, for example, choose the designation 'Hispano' rather than that of 'Mexican-American' (Turner, 1990).

Working in 'the System'

Black and Hispanic leaders have campaigned in local elections to redress the grievances felt by their communities. Many have referred to black or Hispanic 'power' in city politics and have extolled the virtues of working 'within the system' to increase the benefits arising from cooperation with city authorities and the corporate private sector. Hispanic organisations have been active in Washington DC in campaigning on national issues, conducting extensive research and attempting to expand their influence in government. Organisations such as the National Congress for Puerto Rican Rights, the National Council of La Raza ('La Raza'), the National Puerto Rican Coalition and the Cuban-American National Council have been active in the capital (Turner, 1990). La Raza has developed an agenda to campaign for improved housing, employment and healthcare, although these demands are not always realisable where Latinos are in conflict with other ethnic groups such as blacks.

In terms of representation, from none in 1960, by 1985 there were twenty-seven black and three Hispanic mayors of cities with populations over 50,000 (Browning and Marshall, 1986). Despite the continued proportional under-representation of blacks in American city government and their inability to make an impact in many urban areas (Browning and Marshall, 1986), there was a strong impulse for blacks to join with other minorities in biracial or multi-racial coalitions, especially where there appeared to be a strong possibility of blacks and other minorities being

173

able to make an impact on local affairs (Browning, Marshall and Tabb, 1986).

Multi-Ethnic Coalition Politics

The increased representation of American blacks on city councils thus characterised the mobilisation of minorities in the cities and the willingness of community leaders to enter into multi-group coalitions in order to enhance their influence. Where 'atomisation' served to reduce the bargaining power of minority groups, coalition politics sought to enhance the position of organisations and individuals through a form of collective political effort. The sophistication of this multi-ethnic coalition-building provides a contrast with black and Asian groups in the UK where the tendency has been for leading groups to focus their activities around acceptable candidates put forward by the white-dominated Labour party.

The situation in the USA presented blacks with acute difficulties. Even so, there were enhanced opportunities for gaining representation after the 1960s civil rights campaigns. The racist element in US politics persisted as is evident in local election campaigns and in the day-to-day affairs of the cities. In many cases, racism was of a more direct kind than that inherent in UK political life. On the other hand, there were opportunities available to blacks and other minorities in the USA arising from the openness of the political system and its stress upon the importance of elected office at the lower levels of city affairs. Black candidates had numerous opportunities for working themselves up the political ladder by way of lower level elected political office and through party 'machines' that were responsive to coalition-building politicians.

In 1990, the American journal *Political Science and Politics* (PS) published a collection of symposium articles on black electoral politics which indicated the importance of coalition-building. In the introduction to the symposium, Perry (1990a) suggests that the political system in the USA proved to be amenable to black penetration while the American economic system remained difficult to enter. He argues that blacks became a political community; partly confirmed by the fact that in 1986 the voter registration rate of blacks equalled that of whites for the first time in US history. Blacks developed a marked disposition to participate in community politics and local organisations which was notable in view of their low socio-economic status.

This, for Perry, helps to explain the continued advance of blacks in the political life of the USA, again illustrated by the success of minority candidates in elections between 1989 and 1990. During this period there were significant victories for blacks which continued a trend evident in previous years where blacks had gained high office often against strong opposition from white candidates. For Perry, four victories in the 1989–90 period were of particular note. These were the successes of Norman Rice in Seattle, David Dinkins in New York City, John Daniels in New Haven and Douglas Wilder in Virginia, all elected as the first black chief executive in their respective jurisdictions. The Wilder victory was of special historical significance in that he was the first black to be elected as governor in the USA.

The PS symposium articles were concerned about the effect on voting behaviour of such black candidates running against white candidates. It was found that black voters gave near unanimous support to the black candidates and turned out to vote at an equal or even higher rate than white voters. This was the case in the election of Dinkins in New York City and Daniels in New Haven. It was also found that the prospect of a black victory did not result in white counter-mobilisations (Perry, 1990a, p.141), a characteristic of campaigns which has been confirmed in previous research (Perry, 1990b).

The Election of John Daniels as Mayor of New Haven

Summers and Klinker (1990), writing in the PS symposium, cover the 1989 Daniels victory in New Haven. Daniels won 68 per cent of the vote in a city that is only one-third black. The Daniels campaign was remarkable in that it represented an insurgent offensive against Ben DiLieto, the powerful and apparently popular Democratic incumbent. Although Daniels was the most senior and most respected Democratic black politician in New Haven, no black elected officials or black church ministers appeared on the podium at his announcement. Summers and Klinker indicate that his support came mainly from 'a handful' of black party activists, businessmen, students, community organisers, board of education employees and Yale University professors and students 'in a town not known for successful challenges to machine politics or its love for Yale' (Summers and Klinker, 1990, p.142).

The Daniels campaign was well organised and highly professional. Careful attention was paid to creating the right media image for the candidate by appealing to all racial and ethnic groups. Endorsements ranged from a black high school student to a Hispanic secretary, to a Nobel Prize winner. The support for Daniels thus represented a coalition that looked to present a viable alternative to DiLieto. It gave expression to grievances that had been building up for many years and which could now 'go public'.

This worked in Daniel's favour in the local Democratic town committee and eventually led DiLieto to announce that he would not run for re-election. Other contenders had little chance against Daniels as he took the opportunity to consolidate his winning coalition. This coalition developed an alliance between black Democratic party activists and grass roots organisers eager to involve young people 'in mobilising the black community to assert its political power' (Summers and Klinker, 1990, p.143). The mobilisation of support for Daniels was based on demands to deal with the increasing problems of crime, poverty, homelessness, Aids and infant mortality; all issues which the DiLieto regime had been unable to confront in an effective manner, and all matters of concern to a range of ethnic groups. The Daniels campaign was able to claim the moral high-ground and appeal to liberal supporters opposed to DiLieto's admitted wire-tapping of anti-war and community activists in the 1960s.

The David Dinkins Victory in New York City

Thompson's (1990) analysis of the Dinkins victory concentrates on the decline of the Democratic party organisation in the city and the general strengthening of black politics. Two factors laid the foundation for Dinkins' success. The first was his winning of the Manhattan Borough presidency, making him the highest ranking minority official in the city. The second was the weakening of county Democratic organisations in the Bronx and Queens following a major corruption scandal. This created the opportunity for 'machine' black politicians to oppose Ed Koch, the Democratic machine candidate for mayor (Thompson, 1990, p.145).

Dinkins won the Manhattan Borough presidency in 1985 on the basis of a broad multi-racial and multi-ethnic coalition including substantial Jewish and Latino support. Dinkins became the only non-white member of the city's powerful Board of Estimate and the city's most important black politician. The criminal investigation and suicide of the Queens Democratic County Leader/Borough President, Donald Manes, in 1986, together with police investigations and indictments against Democratic political leaders in the Bronx and Queens hastened the demise of the party organisation in two areas that were supportive of Mayor Ed Koch. Black and Latino politicians in the Bronx and Queens took advantage of the situation and enhanced the position of Dinkins there. Latino politicians sought alliances with blacks to win seats left vacant by the criminal convictions. This meant that by '1988 New York City had a changed political landscape. The city's corruption scandals made Mayor Koch increasingly unpopular, and traditionally loyal black and Latin machine politicians were breaking ranks' (Thompson, 1990, p.146). Thompson describes how the Dinkins victory was preceded by Jesse Jackson's attempts to win labour and Latino support. The 1988 Jackson Presidential campaign had been a testing ground for the 1989 mayoral candidacy with Koch viewing Jackson as a direct challenge to his position in New York. Koch did all that he could to ensure a high voter turnout against Jackson. Jackson's victory in the 1988 Presidential primary election in New York thus indicated the emergence of the black–Latin–labour alliance as a power to reckon with despite the problems that had been encountered when Jackson angered

Jewish voters by his less than sympathetic remarks about their role in New York politics. By denouncing the anti-Jewish black minister Louis Farakhan, Dinkins courted the Jewish community whose support was vital in the mayoral campaign.

Dinkins was able effectively to assess the public mood with respect to a growing tide of racism in the city. This gave rise to increased grass roots mobilisation around the race issue and the controversial role of Ed Koch in handling race-related questions. Koch infuriated blacks on a number of occasions by seeming to side with whites on matters concerning violence against blacks and racism in general. Indeed, the race issue was one which was evident throughout the Dinkins campaign. It was defined in terms not only of white antagonism towards blacks, but also of animosities between Jews and blacks.

Dinkins was successful because he recognised that, as an insurgent black candidate, he needed a liberal multi-racial coalition to win. He did not campaign exclusively or even primarily in the black community, but concentrated on the various elements of the white vote. Black voters turned out heavily for Dinkins despite his concentration of effort elsewhere. Ninety one per cent of blacks supported him in the mayoral contest against Republican Rudolph Guliani, along with 65 per cent of Latinos, 18 per cent of white Catholics, 35 per cent of Jews and 14 per cent of other white voters.

MINORITY POLITICAL ACTION: USA

Coalition-building between ethnic minority communities in elections was evident at a time of competition between different minority communities. How could this apparent contradiction be explained?

In the USA, as in Britain, ethnic minority organisations, in the 1980s, existed in an environment characterised by the strength of conservative political forces and by a diversity of urban communities of different racial and ethnic composition. In the USA, these communities tended to be more segregated along racial and ethnic lines, resulting in less of a multi-cultural mix within specific geographical areas. The competitive element became of particular significance as local organisations struggled to maximise their resources. It was the lack of unity between different ethnic minority groups which tended to undermine their collective potential. This reinforced the tendency for group leaders to want to compromise with local governments and politicians and maximise the benefits bestowed by the political system. Cooperation was an effective way of achieving this.

In the USA, the 1960s and 1970s was a period of division within the black community. The rise of black nationalism and the influence of organisations such as the Black Panther party were not aberrations. They represented the expression of a desire by many blacks to break from the

177

confines of what they regarded as being a white racist state, and, to some extent, they represented a 'left wing' in the broader campaign for civil rights. This current provided a means by which radical black organisations could organise against moderate leaderships which were seeking to bring about reforms within the prevailing democratic political system.

In Britain, the left wing in the black community remained weak and poorly organised. In the USA during the 1960s and early 1970s the black radicals actually began to undermine moderate leadership roles in certain black communities and were in a position to make important political impacts during and after communal disturbances in major US cities. The 'community power' perspective adopted by many black radicals thus appeared to represent a practical programme for change in many black ghettos and an alternative strategy to the reform proposals of moderate office-seeking 'professional' leaders.

Ironically, the appeal of the 'community power' position provided the basis for the political defeat of black nationalism. The Black Panthers regarded their influence in defensive terms, and claimed that community power was a means whereby blacks could defend their civil and legal rights under the US Constitution. This demanded the development of community defence through legal channels, the extension of community participation in local projects and the lobbying of local officials and politicians (Seale, 1970). Community 'power' was converted into community 'control' and 'participation' which were terms then in fashion with liberal Democrats and moderate white community workers. As in Britain, the practical questions facing blacks tended to strengthen the position of the moderates and encourage radicals to orientate towards well-established organisations and agencies (Baraka, 1972; Gilliam, 1975; Perkins, 1975; Hillson, 1977).

The US Response to Welfare Liberalism

The accommodation of black politics in the USA was accompanied by the development of a conservative black politics which reflected the degree to which some blacks had come to identify with the values of mainstream politics. The conservative notion of decentralised government intersected the radical/liberal desire for community control. The assertion of the need for local control became attractive to conservatives during the late 1970s and early 1980s when control was linked to an economic critique which stressed that the welfare-liberal approach to providing community services had failed.

In the USA, even during the latter part of the Reagan era, there was a feeling amongst conservatives that interventionist programmes were acting as a constraint upon distressed communities by reinforcing the economic divisions between blacks and whites. American conservatives claimed that the market was the way to salvation for ethnic minority groups and deprived 'inner-city' whites. The emergence of a new conservative agenda for blacks in the late 1970s and 1980s marked both a challenge to the left and a confirmation of the assertions that the conservative argument acted as a radical alternative to the allegedly discredited policies of welfare-liberals (Sowell, 1975). In the 1980s, it was the conservative agenda which conservatives claimed to be gaining momentum at the expense of liberal notions of reliance on the state and the quest for greater social equality. It was private enterprise which promised to provide the more dynamic solution and which held the prospect of raising a new black entrepreneurial class from the deprivation of the ghettos (Bolick, 1988).

This view was taken by the Washington DC based Heritage Foundation. The Foundation was established in 1973 as a non-partisan, non-profit research institute dedicated to the principles of free competitive enterprise, limited government, individual liberty and a strong national defence. Conservative policies, including a free market approach to regenerating the cities were advocated. The Heritage Foundation was concerned to win black Americans to the conservative banner despite their more traditional Democratic orientation. In a collection of writings published by the Foundation in 1987, entitled *A Conservative Agenda for Black Americans*, it was asserted that

> Conservatives have built a movement that is transforming America's political landscape. To be of lasting importance and historic proportion, this movement must encompass all Americans. Thus, conservatives must convince blacks of the virtue of the conservative ethos.
>
> Bringing blacks into the conservative movement is not impossible. The challenge, in fact, becomes not all that daunting, once conservatives realise that the black community is not monolithic and that those who are cast as spokesmen for blacks are often out of step with the masses of blacks. This was the message of a much discussed 1985 survey by the Center for Media and Public Policy, a nonpartisan, privately funded research group. For the first time, the survey compared the political attitudes of blacks and their leaders. Among other things, it revealed that the majority of blacks feel that those claiming black leadership have no mandate from the masses of blacks. It also showed that grass-roots blacks differ with their putative leaders on a wide range of issues. Yet, to the frustration of conservatives, black Americans continue to vote for liberals, who are completely out of step with their views on many of the most basic issues.
>
> (Perkins, 1987, pp.1–2)

The 1988 Democratic Presidential nomination campaign of Jesse Jackson was widely 'out of step' with the desire of conservatives to develop effective market economic initiatives. Jackson mobilised a substantial number of black Americans around a programme of demands which were liberal on questions concerning the fate of the ghettos. However, Jackson was a strong advocate of private capitalism and entrepreneurialism in the ghettos as a way out of deprivation and unemployment. In that respect, the Jackson campaign embodied many of the market prescriptions of those on the right. His campaign arguably represented the incorporation of aspects of the conservative agenda into the political demands of America's 'mainstream' black leadership.

ETHNIC MINORITIES IN BRITAIN

There were similar attempts to encourage blacks and Asians in the UK to follow the entrepreneurial road, although the intellectual climate was not so favourable to British black conservatives. The Major and Thatcher governments were supportive of private initiatives. Organisations such as Business in the Community promoted minority projects involving the community and business. There were even interchanges between British and American community activists involved in black business development (Community Service Volunteers, 1985). However, UK traditions differed.

'Race Relations' and Community 'Harmony'

In Britain the formation of government bodies to promote the political integration of diverse black communities has been important since it represents an attempt by government to overcome some of the problems which arose following the 1950s and 1960s periods of black immigration. It was an attempt to implement a 'race relations' policy which involved a measure of consultation and limited 'participation' for black people as a whole (Rex and Tomlinson, 1979; Edwards and Batley, 1978). An integral part of this approach in the 1960s and 1970s was the establishment of community relations councils and the government-funded Commission for Racial Equality.

Critics of initiatives to bring about community 'harmony' by way of government intervention pointed to a duality in government policy. While there were attempts to promote 'good community relations', account had to be taken of the abrasive affects of a rigorously applied immigration policy and so-called 'physical measures' used to control riots and other social disorders (Humphry and Ward, 1974). The combined outcomes of physical control and immigration restriction made black leaders suspicious of the central government in its application of community development policies. This suspicion extended deep into many black communities where the experience of ghetto life and of alleged police 'harassment' led to a loss of confidence in public authorities in general (Scarman, 1981).

In the USA blacks were integrated into a range of community activities, entered partnerships with local governments and developed highly professional representative organisations (Williams, 1982). This process was less thorough in Britain. In the USA community programmes were initially provided with levels of funding which produced some incentive for even quite radical blacks to cooperate in local development projects (Marable, 1985). In Britain political participation was achieved on a comparatively modest scale where groups had to operate within comparatively 'underfunded' community projects attracting only modest private sector support (Jacobs, 1986).

Black and Asian leaders in Britain frequently played a balancing role by both presenting moderate demands to politicians and government officials and also developing populist or radical demands to echo community discontents. This was characteristic of the demands of many secular leftist political groups arising from clashes between the police and black and Asian groups during the 1970s (NCCL, 1980). For example, this is the case with the left-inclined Indian Workers' Associations (DeWitt, 1969; Beetham, 1970).

MINORITY REPRESENTATION IN THE UK

Despite the community relations movement, in political terms recently the more important debate about black and Asian advancement in the UK has focused upon representation. It is noteworthy that the debate in the UK was central to demands to improve the position of non-whites in British political life. The electoralist orientation of groups calling for reforms to effect greater black and Asian representation was a feature of minority politics in the UK during the 1980s, particularly as minority candidates gained office and were elected to Parliament. Yet, there remains a clear under-representation of non-white minorities in Parliament when measured in proportionate terms against the size of non-white communities.

Racism in the System

Recent disputes within the Labour and Conservative parties about alleged racism highlighted the problems faced by non-white minorities in gaining representation in official positions within the parties and in gaining seats in Parliament. The desire by the British political parties not to upset the status quo on race issues was strengthened by the long-held assumption that there is 'neutrality' in the handling of minority representation in British politics (Gudgin and Taylor, 1979). The notion that the distribution of seats in national and local elections is governed by purely organisational as opposed to political considerations is extremely strong, as is the idea that the political process enables the expression of minority interests though a racially neutral democracy (Jacobs, 1989a).

There have been attempts to indicate that the 'neutrality' of the system may in fact cover a strong political bias, in politics in general and in the electoral process in particular, which effectively produces discrimination against ethnic minorities. This bias seems to permeate the system in a variety of ways at both the national and local levels. Electoral arrangements may be regarded as operating within a broader context in which there is a pervasive racism, conditioned by strong attitudinal dispositions which work to the detriment of minority candidates in elections and in other ways in the representative political system.

It is the notion of 'neutrality' in the British system which helps to preserve the bias in such a way as to reduce political debate about its inherently discriminatory aspects. Political controversy about race-related issues is set within this constraint despite the fact that many minority organisations have raised criticisms of the political parties' failure to promote minority representation. It is evident that the more political and partisan characteristics of other democratic systems (such as in the USA) are more conducive to genuinely open discussion about the racist aspects of their electoral systems (concerning open discrimination, party bias and electoral boundary issues, for example). This assumes that within an open political system minorities are to be afforded the opportunity to gain access to both political and economic power centres (Glazer and Young, 1983). It also implies that there will be open debate about all the racist characteristics of the system so that they may be overcome or changed.

Modest Successes

The low-key attitude towards such issues in the UK was characteristic of public debate in the 1987 general election which saw the election of four non-white members to Parliament (all members of the Labour party). These were the first non-whites to be elected in nearly sixty years. Diane Abbot was returned for the Hackney North and Stoke Newington constituency in north London. Paul Boateng was elected in Brent South in west London, Bernie Grant in Tottenham in north London and Keith Vaz in Leicester East. These successes were accompanied by an increasing number of black and Asian candidates standing for Parliament since the 1979 general election. The number of ethnic minority Parliamentary candidates from the four main political parties rose from five in 1979 to twenty-two in the 1983 general election. In 1987, the figure increased to twenty-nine minority candidates from a total of 633 Parliamentary seats.

Le Lohe (1989) provides the most detailed account of ethnic minority performance in the 1987 general election. His study indicates that twenty-six of the twenty-nine candidates were placed in constituencies where their

182

own political parties had not expected to win. These were regarded as marginal or 'hopeless' constituencies. The Labour party had been more willing to adopt non-white candidates in relatively 'good' seats such as those won by the three black MPs in London. The former Alliance parties (the Liberals and Social Democrats) offered none of their potentially winnable seats to non-white candidates. There was a similar, but not quite so bad situation in the Conservative party.

At the local level, there was an increase in the number of non-white candidates standing for election since 1979. However, despite increased ethnic minority representation, major difficulties remain for non-white candidates seeking office. Anwar (1986), using data collected by Le Lohe, points to the continued under-representation of non-whites in local elections as a feature of local politics in Britain going back to the 1960s when there was significant black and Asian immigration into Britain. In the 1981 county elections, there were fifty-four ethnic minority candidates of whom only twelve were elected. This constituted only a tiny proportion of the 4,000 county councillors in England. At the 1982 local elections in Birmingham there were 375 candidates of which only twenty-three were of ethnic minority backgrounds. Only five ethnic minority candidates were elected in a city where 15 per cent of its population was non-white.

As in the USA, local situations vary from city to city, and so it is difficult to make generalisations about the nature of minority political strategies and alignments that hold universally. However, as in the case of elections to Parliament, blacks and Asians in British local electoral contests have most frequently joined together in voting for candidates which have claimed to represent all minority groups within particular communities and this has tended to strengthen the notion of a 'black identity'. In the early 1990s there was little evidence that this pattern was changing radically in spite of the growth of a variety of radical and fundamentalist political organisations committed to various brands of ethnic, racial or religious segregation.

Some political organisations preferred their own candidates where there was a strong ethnic identity. Sometimes this led to challenges to the main parties when ethnic minority groups stood their own candidates (such as in the 1989 Vauxhall Parliamentary by-election in London). Generally, such challenges have threatened to pose the greatest danger to the Labour party, since it is through the local branches of the party that so many ethnic minority interests find expression. There have been instances when Asian and other interest groups have operated directly through Labour party branches to influence candidate selection and party policy (Jacobs, 1986). Such penetration of the Labour party has frequently produced internal disputes about the legitimacy of the actions of inner-party pressure groups.

In London the rise of the new urban left black local council leaders such as Bernie Grant in Haringey, Merle Amory in Brent and Linda Bellos in Lambeth, also provoked a wider debate within the Labour party about the role of blacks in British political life and about the provisions that should be made for ethnic minorities within the structures of the political parties and in local government. Much of the debate in the Labour party was initiated by the left's unofficial party Black Sections movement which campaigned for increased black and Asian representation at the local and national levels and for the democratisation of the Labour party to achieve enhanced minority participation and membership.

COOPERATION, CONFLICT AND DISSENT

Difficulties in gaining adequate representation produced frustration within black and Asian communities about their role in the British political system. There have even been calls by some Islamic organisations for separate political institutions to handle their demands. However, there are wide differences of opinion within minority communities in the UK as to how blacks and Asians should strive to advance their democratic rights and how they should relate to prevailing political institutions.

Unlike the USA, there are no major national 'peak' black representative interest groups (such as the US Urban League or the National Association for the Advancement of Coloured People) in the UK. If there is any kind of common focus for black politics at national level in Britain it is provided by the Labour party. But even the Labour party finds it hard fully to play such a role when blacks themselves remain organised largely in small localised groups. In addition, the Labour party leadership has resisted the formation of special party 'black sections' in order to maintain party cohesion and to move away from the idea of separate national black political mobilisation.

This situation suits many moderate black leaders who are keen to maintain their own political influence. A stronger national black identity could threaten their local bases of power and their hold over the smaller ethnic-based organisations. The concept of a national black identity is therefore frequently associated with the demands of the political militants who see a national organisation as being a prerequisite to the mass political action of working class blacks to achieve social and political liberation within what they regard to be a white racist society. More conservative local leaders, who do not share this view, see the radical demands of the militants as a dangerous threat to the cohesion of traditional communities and as a way of undermining old 'homeland' customs and religions.

The issue of a national black or multi-ethnic organisation serves as one illustration of the character of mainstream black leadership in Britain. It also highlights the fact that leadership perceptions relate to factors concerning subjective assessments of the costs and benefits of interest group strategies under conditions which demand awareness of conflicting political influences (Jacobs, 1986, 1988). Traditional black leaderships are well placed within their own communities, since religious and cultural practices tend to be solidly entrenched amongst the 'rank and file' followers of organised religious and other interest groups. The support for traditional values may indeed have been strengthened to some extent by the rise of fundamentalist ideologies abroad and the willingness of the younger generation of blacks to follow various brands of homeland nationalism and militant Islamic ideology.

The controversy over the 1989 Iranian call to eliminate the author Salman Rushdie in response to his book *The Satanic Verses* forced a division between Islamic fundamentalists and moderates in the Asian community. The Rushdie issue became prominent at a time when Islamic fundamentalism was already gaining ground in Britain and as British Muslims were developing a stronger sense of their own identity in response to racism. The 1991 Gulf War also posed important questions for British Muslims about their attitude towards Britain and the British government in view of its involvement in the campaign to remove Iraq from Kuwait and wage a wider war (Khanum, 1991).

Moderate Leaders

Strong ideological and religious commitments tended to cut across class identifications and make it more difficult for black and Asian militants in Britain to campaign on basic trade union and labour related issues. These issues usually form the basis of radical socialist and communist political campaigns waged in different localities, particularly during periods of communal disorder when the militants calculate that traditional allegiances are most prone to breakdown.

Established community leaderships are thus placed in a difficult position during times when social tensions are near the surface. Social instability, combined with the more 'normal' problems faced by blacks and Asians in economic terms, provides community leaders with a challenge to maintain the ordered channels of communication with local government and politicians that are necessary to ensure that access to public policy-makers is not disrupted.

Moderate leaders are also aware of the potential power of populist appeals for 'black unity'. A common feature of the post-riot situations of the 1980s was where moderate community leaders reflected both the traditional and more militant sounding demands of black youth. Moderates were successful in holding on to community support for policies which supported cooperation with white politicians, local council officials and with the police. They echoed some of the demands of militants within the black and Asian communities, but within the context of a policy of co-operation and support for racial and communal harmony (NCCL, 1980).

RIOTS AND LEADERSHIP

Similarly, in the USA the moderate response is seen to be supportive of 'harmonious relations'. This is not always easy to achieve. The 1980 disturbances in Miami left eighteen dead and large areas of the Liberty City district affected by violence and arson. Blacks felt threatened by Mexican and Cuban immigrants at a time when federal cutbacks choked off funds for urban programmes (Neil, 1982).

A November 1982 demonstration in Washington DC had more of a political character. It arose following clashes between the police and demonstrators at an anti-Ku Klux Klan march. A riot developed, resulting in the arrest of thirty-eight people and damage to property.

In the Miami riots of 1989, in the Overtown and Liberty City neighbourhoods, moderate politicians in the black and Hispanic communities were concerned to blunt the impact of inflammatory statements made by elements which sought to create divisions between ethnic minority communities. The moderates had to 'play' to their own communities while attending to the task of calming community tensions. Such tensions were exacerbated by the increasingly racial nature of politics in Miami and Florida as a whole as exemplified by the comments of the Democratic politician, Gerald Richman who was alleged to have played upon anti-Hispanic sentiments in his 1989 state election campaign.

The moderating attempts to maintain order and support legitimate participatory channels were reminiscent of the leadership reactions in the USA following the riots of the 1960s and 1970s. However, in the USA in the 1960s, the established moderate leaders faced well-organised opposition groups such as the Black Panthers which had a national reputation and enjoyed some support within black communities. In the UK, the 1981 and 1985 riots emphasised the relatively weak position of radical/leftist organisations in the community (a situation which was also characteristic of the Miami disturbances in the 1980s).

The Balancing Role of Leadership

The task of moderates in those kinds of situations was generally to restore 'normal conditions'. This is not only regarded by the moderates as a preliminary for continued access to government, but also as a condition for consultation on matters concerning the future of urban policy in the localities affected. Therefore, as in the USA, moderate community leaders were an important part of the process of restoration and an essential support for the development of programmes to assist with the development of local economies and social infrastructures.

The 'balancing' role of community leaders can be drawn from the following experiences in UK towns and cities.

Brixton, 1981

The disturbances which occurred in the Brixton district of south London between 10 and 12 April 1981 represented a potentially disruptive situation for established community leaders. In their intensity, the 'riots' went far beyond those experienced in Bristol in 1980.

The disturbances were the subject of an official government inquiry headed by Lord Chief Justice Scarman (Scarman, 1981). The Scarman report, published after a detailed investigation into the causes of the riots, is particularly illuminating when it describes the role of community leaders. Scarman clearly indicates that the representatives of the black community in Brixton were keen to play a mediating role between the rioters and the police in order to bring a speedy end to the disturbances. This was in spite of the evident anger of the crowds gathered in the streets and the implementation by the police of the highly unpopular 'Operation Swamp' as a measure designed to combat local crime. Community leaders met with the police on several occasions as instances of looting and petrol bombing were reported.

After the riots, the main concern of moderate leaders was to restore confidence in legitimate channels of communication between the police, local people and the local authority. In spite of a poor history of police–community relations, moderates saw that there were benefits to be gained from a policy of cooperation, especially as it was important for Brixton to remain attractive as a place for companies to invest and provide jobs through economic growth.

The 1981 Summer Riots

A similar leadership response was evident in the summer riots of 1981 where it was again disorganised youths who took to the streets without any clear conception as to where their actions would lead. Disturbances occurred in July with serious outbreaks of street violence in many towns and cities beginning with Southall in London and Toxteth in Liverpool. There were major disorders in Handsworth in Birmingham and in many other towns and cities throughout the country. The trouble in Toxteth continued intermittently until 6 July, during which time the Merseyside police were reinforced by officers from surrounding constabularies and the use of CS gas was sanctioned to disperse the crowds.

The police maintained that the riots were not 'race riots' since there were blacks and whites involved. This was a point that was made by the government. This helped to ease the way for a visit to Liverpool by the Prime Minister on 13 July to talk to community and church leaders. There was some suspicion of the Prime Minister's motivation so many black and white local activists opposed meeting Mrs Thatcher. However, opposition to this liaison was surprisingly muted since even some radical organisations saw that they had little to gain from the disruption caused by the riots. Many who felt that there should be a policy of non-cooperation with the Scarman inquiry were concerned that there had to be a dialogue with local politicians and government officials so that the community could restore confidence in itself and project a plan of action for the future (Jacobs, 1986). Local councillors were in agreement with a policy of cooperation with the government because there was the prospect of extra government funding for programmes in Liverpool following directly from the consequences of the disturbances. Indeed, Liverpool was selected as a priority area by the government in its post-riot urban initiatives and regarded as a test-bed for new ideas on how to revitalise the inner cities.

In Manchester, following the disturbances in the Moss Side district, local community leaders believed that the Scarman inquiry could provide them with an opportunity of making representations to an official inquiry on matters of deep concern to local people. The Greater Manchester Council (which was later abolished as part of Mrs Thatcher's campaign against the urban left and the metropolitan councils) established its own panel of inquiry into the disturbances chaired by Brent Hytner QC. Despite the fact that the panel had no statutory powers to call witnesses, some militant organisations decided to boycott its deliberations. However, the panel did receive the support of more moderate groups and even produced a report which was critical of some police measures that had been implemented in Moss Side.

The 1985 Riots

Many new urban initiatives followed the 1981 disorders (Clark and Rughani, 1983; Brooks, 1983; Dungate, 1984; CRE, 1985). They were based upon the cooperation between moderate leaders and public and private agencies of various types (see Chapter 9). Despite this there were further disturbances in major British cities in 1985.

On the night of 6/7 October 1985 Britain witnessed the most violent scenes ever in a mainland disturbance. The disorder resulted in the death of a policemen (Police Constable Blakelock) and the injury of large numbers of youths and police officers. The disorders, which took place on the large Broadwater Farm housing complex in Tottenham in north London, followed serious riots in Handsworth in September and in Brixton and Toxteth in October.

These disturbances were prefaced by a number of incidents involving the police and mainly black and Asian youths which led to criticisms by a range of organisations about the way in which the police were handling relations with ethnic minorities. The continued problems in this field often made it difficult for moderate organisations to maintain their cooperative stance with respect to the police. As in the USA, even non-militant black leaders responded by being critical of measures which they regarded as discriminatory or racist. The constant tension between operating 'within the system' and the need to respond to grass roots demands about racist situations often posed a problem for those leaders who desired a relatively ordered political environment. Both during and after the 1985 riots, organisations such as the Indian Workers' Association, religious groups and community advocacy organisations increasingly found themselves outflanked in their populist demands by leftist–radical and religious fundamentalist groups which directly campaigned against established political institutions and processes of consultation.

In Handsworth in Birmingham, during the 1985 riots, there was some evidence of tension between moderate Asian leaders and moderate Afro-Caribbeans following the destruction of commercial properties by black and white youths. The conflict was reflected in the local Community Relations Council which acted as the local government-funded body established to promote better community relations in Birmingham. Dissent between conservative community leaders provided militants with the opportunity of intervening with their radical political programmes in a setting which was highly unstable (Jacobs, 1986). Traditional religious leaders were concerned to maintain order, but were also prepared to undermine militants by concentrating on sectarian community issues. This kind of confused alignment of forces may well contain the seeds of a serious breakdown in future communal order, but in the Handsworth case it appears to have served the prevailing community notables well in that they successfully averted the short-term growth of any major organised black political opposition in Birmingham.

Sectarianism may, therefore, be regarded as a serious impediment to the development of a major national black organisation, but not necessarily of long-term political benefit to conservative leaderships. As suggested above, narrow sectarianism often produces conflict and disharmony. Not surprisingly, there are many black moderates who would now advocate greater cross-cultural and cross-racial unity to achieve an effective national black lobby.

LEADERSHIP IN CRISIS

The question arises, therefore, as to what should be the objectives of ethnic minority leaders in Britain. Are there important lessons to be learned from the USA? In the USA, there was public dissatisfaction with the record of some blacks in office and disappointment and frustration at the way some black officials behaved once in office. Disillusionment opened a debate about the various strategies open to minority groups and a reassessment of approaches to minority empowerment. As Browning and Marshall ask

> How much power do black and Hispanic officials have? And what difference does it make in any case? Can blacks and Hispanics in office strike a better bargain with investors and financial institutions they depend on for development than previous administrations? In a period of fiscal stringency and reduced federal spending, can they carve out a larger piece of the municipal pie for minority populations? Can minority officeholders make any headway against the growing tide of unemployment and poverty, as help from the federal level is dismantled? Although the growing number of minority officials suggests political integration and racial–ethnic succession, do the limited powers of cities in a federal system and a capitalistic society render that apparent gain more symbolic than real?
> (Browning and Marshall, 1986, p.573)

'Cracking' the Image

The case of the former Mayor of Washington DC provided a tragic example of the failure to live up to high expectations. Black Mayor Marion Barry was brought to trial in 1990 to face criminal charges, including possession of cocaine, and felony relating to allegations of lying under oath to a grand jury. The case followed revelations in 1988 that Barry had been associated with a former city employee, Charles Lewis, who was convicted in the Virgin Islands on drugs charges in 1989. Before a grand jury investigation Barry initially denied that he had taken drugs. In November 1989, Lewis testified that he had given 'crack' to the Mayor on more than one occasion.

At the trial in August 1990, the jury consisted of ten blacks and two whites. They reached a consensus on only two out of fourteen counts. Barry was convicted on one count of cocaine possession and acquitted on another. Judge Thomas Penfield declared a mistrial on the twelve other 'deadlocked' counts. Shortly before the trial, Barry had announced that he would not seek re-election as Mayor. Several of the jurors openly expressed concerns about the quality of the evidence that had been presented at the trial. They were of the opinion that Barry had been 'set up' as a way of removing a black from office (*The Washington Post*, 23 August 1990).

The concerns about the racial implications of the case were expressed by both blacks and whites. Barry stressed the relation between what he regarded as personal attacks and the alleged racial nature of the prosecution case against him. There was talk amongst his supporters of a 'white conspiracy' to 'get' the Mayor and allegations that the FBI had encouraged a smear campaign. While probably there were opponents of Barry who would have liked to have seen him removed on racial grounds, the issue was partially defused by the intervention of the black politician, Sharon Pratt Dixon, who had called for Barry's resignation. In the event, she won the Democratic nomination and was later elected as mayor. Dixon's nomination was popular with both middle- and working-class blacks and whites.

In October 1990, Barry was jailed just six days before seeking election to the city council. There was continued backing for him within the black community and among prominent black politicians who supported his view that the case had racial implications. (Riddell, 1990b). Nevertheless, the jailing of a such an important black politician on charges connected with drugs was a worry to police and social workers working in communities affected by the drugs problem.

Bradley Verses Gates

In Los Angeles, the issue which put the credibility of black Mayor Tom Bradley on the line was that of police brutality. A month after the televised videotaped pictures of the beating of a black motorist by officers from the LA Police Department (LAPD), Bradley demanded the resignation of LAPD Chief Daryl F. Gates. Gates refused to go despite his suspension by the police commission. The city council backed Gates and forced the Mayor and the commission to back down.

The issue had serious repercussions for race relations in Los Angeles and was an issue in the April 1991 city council elections. Gates, with his strongly held views, added fuel to the controversy by firmly defending his department and by claiming that he had been 'disgraced and defamed'. Gates claimed that the issue had raised important questions about policing and crime and that 'something good' would come from public discussion of the issues at stake.

The case concerned the relationship between the Mayor and the council. Bradley was in a weak position. According to his critics he seemed to be ineffective in dealing with an important issue which had profound social and political implications for blacks and other minorities in Los Angeles and elsewhere (a federal review of allegations about police brutality was announced following the LA incident and reports about racial problems, particularly in New York and Chicago). In July 1991 Gates announced his retirement following the report of an independent commission headed by Warren Christopher (a former US deputy secretary of state) which concluded that Gates had tolerated racism in the LAPD and that he should therefore make way for a successor.

Riots in Washington DC, 1991

The police brutality issue was again of central concern following the disorders that occurred in Washington DC in May 1991. The disorders resulted in the imposition of a curfew in the Mount Pleasant district (about two miles from the White House) and in parts of the Adams-Morgan area. The scenes resembled those in the 1981 and 1985 riots in the UK, with local stores and vehicles set alight and police in riot gear patrolling the streets.

Those involved were predominantly Central American immigrants responding to an incident in which the police shot and wounded a Salvadorian man who they claimed had been carrying a knife. The disorders involved angry youths who confronted the police and spilled over 16th Street near to the city's central commercial district. There were reports that many local residents had complained to Mayor Sharon Pratt Dixon claiming that the police had adopted heavy-handed tactics in their dealings with the rioters. The demands from the Hispanic community for greater representation of Spanish-speakers in the police force highlighted long-held concerns within the community about alleged discrimination against Hispanics by the police, and, implicitly, by local black politicians.

'Fiscal Gap' Hits Dinkins

As well as charges that there was a mounting problem of racism in the city, it was the economic realities of life that eventually most served to undermine Mayor Dinkins in New York. In 1991 the city faced an acute fiscal crisis as the recession eroded tax revenues and the costs of welfare and services escalated. The Mayor was sharply criticised by opponents such as former Mayor Ed Koch and labour union leaders who accused Dinkins of not having a properly worked out strategy. However, Dinkins presented his 'doomsday budget' in May: a plan that threatened to cut US$1.5 billion through large job cuts and closures of services. This included possible lay-offs for teachers and council employees, the possible closure of Central Park Zoo and the elimination of selected health and social services. The Dinkins strategy was to seek state assistance to 'fill the fiscal gap' and offset the harsh measures proposed. However, the prospects for 1992 were grim.

CONCLUSIONS

The political incorporation of ethnic minority politics, whether partial or comprehensive, was thus no guarantee of economic success or effective political empowerment. In the UK and the USA the potential for social disorder in distressed communities existed. Conflicts between different ethnic and racial groups often lay beneath the surface, worsened by the struggle between groups for scarce resources.

The credibility of some prominent black politicians in the USA had been undermined. In the UK, black members of Parliament were accused of having achieved only limited gains for their constituents and were criticised by groups seeking more radical change.

The political realities of working within legitimate political channels were as significant for ethnic minorities as they were for others. The problem was that the issues associated with the constraints faced by blacks, Asians and other minorities were posed with reference to race and racism. It was racism, combined with the special problems of minorities in society, which made it more difficult for them to feel fully 'at home' within the political system.

9

CORPORATE VISIONS

The concept of 'business leadership' has arisen from the corporate private sector's involvement in urban renewal and the desire by companies and market-orientated public policy-makers to ensure that enterprise is at the heart of urban regeneration. In the USA, the role of business in local affairs has been well established.

CORPORATE POWER

Corporate power has shaped many of the policies affecting social and economic development in America's cities (Bernard, 1990). Friedland (1982) points to the influence of corporations in Pittsburgh – a city which had attracted 21 of the top 500 industrial company headquarters by 1960. This presence gave the corporate private sector a discernable political weight in the city and provided one of the conditions for its noteworthy regeneration in the 1980s (Weber, 1990). Harding (1990) points to the importance of US 'growth coalitions' which often develop a strong 'independent' executive capacity for city development policy and act as powerful lobby groups. Adams (1990) shows how the Philadelphia business elite developed its own demands and strategy after the Second World War through the Greater Philadelphia Movement (GPM).

Harrington (1984) critically examines the corporate presence. He alludes to the role of chambers of commerce in American cities which often operate as political 'powerhouses' and act as a central focus to link various commercial and industrial interests so that they may more effectively articulate the demands of the private sector. Harrington sees this as a challenge to local democracy as local corporate élites are able to exert their power without recourse to the electoral process.

Harrington notes the influence of corporations in Boston where their influence and power became associated with the civic pride of the city. Boston's top business leaders were organised in the Coordinating Committee or 'the Vault' which gave an important stimulus to the city's economy in the 1960s and 1970s. The Vault saw the development of the central business district in Boston as a vital element in the promotion of the city in the US economy and internationally. It was the chairmen or presidents of major financial and commercial institutions which made the Vault a prime mover in Boston politics.

Stone (1986, 1989) is more sympathetic to the corporate private sector. In Atlanta companies have worked with community organisations to the general benefit of the city and local people. The civic culture has been supportive of socially responsible companies and consequently there has been a 'natural' involvement of the private sector in urban regeneration.

The role of business in city politics in Britain has been generally more restrained. However, business does not just stay on the sidelines. In Birmingham (UK), for example, economic change and the development of initiatives promoted by the city council have increasingly involved the private sector. Large companies have been involved in partnerships with the local authority in the central business district and have often come into conflict with interest groups (such as the Birmingham for People Group) over important planning issues.

The decision-making on planning issues has been influenced by property development companies. City 2000, representing 160 member companies in Birmingham, acts as a pressure group to promote commercial and business interests in the city (Smith, 1991). The group has a strong professional membership which complements the work of the local chamber of commerce. The chamber and City 2000 act as lobby organisations in an environment where the city is attracting more international companies and foreign banks. These international corporate interests themselves exert an influence over strategic planning decisions and are operating as a new political factor in the city.

BUSINESS LEADERSHIP AND SOCIAL RESPONSIBILITY

Attempts to develop business leadership in urban renewal are not necessarily accompanied by a major corporate presence in local politics. In the UK, business leadership has been seen more in terms of involvements in specific partnerships where leadership has to be nurtured. In this context, the concept of 'business leadership' relates to the promotion of business-led initiatives in which groups are committed to the promotion of economic and social development in distressed urban areas. It is concerned with the development of the capacities of community groups to link their efforts to those of companies which are willing to provide support in a number of ways including managerial, financial and organisational development at the local level. This explicitly means that public–private partnerships will strive to develop the skills of participating organisations as part of the process of community-building. Community-level leadership is therefore seen as an important element in the process of group professionalisation.

The concept of business leadership is an appealing one for companies especially when it is defined in terms of overcoming the 'apathy' of communities and public sector bureaucrats (CBI, 1988), and when it is combined with the notion that companies can be socially responsible (Cowie, 1985). There may be 'spin-offs' for companies relating to their longer term trading prospects in a community, particularly if partnership initiatives are intended to promote social harmony. It is such appealing qualities which enable partnerships to assume a key role by attracting the support of diverse companies, community organisations and local politicians.

Leadership Groups

Central to effective business involvement is the creation of active leadership groups. These groups can provide a dynamic force behind urban renewal. Public–private partnerships enable companies to identify themselves with social causes which are not necessarily directly perceived by those in local communities as being conducive to traditional profit-making activities. Corporate social responsibility maintains that companies will adopt a positive and supportive attitude in distressed communities, and that they will take a leading role in developing community capacity and resources for economic development.

Partnership is seen as a way to generate socially responsible developments which can provide an attractive alternative to those developed by city hall bureaucrats. The development of leadership potential in the private sector and within voluntary and non-profit organisations is thus seen as an aspect of community invigoration outside of public bureaucracies.

Corporate involvement should therefore stimulate the development of community-level leaderships. Local leaderships will be committed to the objectives defined by public–private partnerships and will be provided with opportunities for developing their own innovative approaches to local problems in consultation with local people and relevant professional organisations and individuals. Cowie (1985) alludes to the central importance of the development of such leadership potential in communities. This new leadership in communities is expected to provide the impetus for change. The initiative has to come from the corporate private sector to aid the development of core groups within communities, and this can be achieved by way of the assistance of company personnel in urban renewal.

The Benefits for Government

Where governments are involved, potential benefits may arise from corporate social responsibility. Governments, particularly at the local level, will usually be keen to attract development to depressed areas. There is an incentive to establish good relations with companies that show an interest in assisting localities to achieve economic expansion and create job opportunities (Jacobs, 1989b). Where business leadership is the accepted mode of operation, there is an assumption that it will be the corporate private sector which is the motive force behind community regeneration even though local government may have an important facilitator role to play.

Partnerships have provided the public and private sectors with new opportunities for encouraging communities to further their interests and coordinate their politics through new structures. The partnership concept can assume many guises. There are many different arrangements which bring together disparate organisations seeking to benefit communities. Partnerships in the USA and Britain have evolved a multiplicity of approaches which make it difficult to generalise about the forms of coordination between governments and other groups.

TYPES OF PARTNERSHIP

One approach to providing a framework within which partnerships can be better understood was that provided by the Massachusetts Governor's Task Force (GTF) on Private Sector Initiatives (1983). The GTF report described partnerships operating in Massachusetts, but it is relevant to describing partnerships in a variety of settings. The criteria for judging the activities of partnerships is useful in characterising public–private interactions and providing decision-makers with guidelines for evaluating their performance. The GTF report indicated the different roles ascribed to community organisations in partnerships and suggested that there was a range of participatory structures that could accompany the formation of a partnership.

The GTF report was a response to a request by the Presidential Task Force on Private Sector Initiatives (which was abolished by the Bush administration). The GTF was charged with the task of investigating the role of the private sector under three main headings, namely, public/private partnerships, charitable finance (including the activities of organisations working 'not for profit') and 'volunteerism'. The task force was chaired by John Larkin Thompson, President of Blue Shield of Massachusetts. It was organised into four working subcommittees to reflect its major concerns. Assistance was given by other organisations including the Boston Private Industry Council, the United Way of Massachusetts Bay (see Chapter 10) and the Lincoln Filene Center for Citizenship and Public Affairs at Tufts University.

A subcommittee on public–private sector partnerships developed an analytical framework and produced a description of 120 partnership projects in Massachusetts. It also produced case studies of six outstanding public/private sector partnerships which detailed the steps taken to form and implement successful partnerships. These case studies were made available through the Boston Private Industry Council as an appendix to the main GTF report.

The GTF report referred to the tradition of private initiative in the USA and Massachusetts. It was America's early history which gave rise to the view that citizens were responsible for taking their own initiatives and were not dependent solely upon government for the provision of services and social support. The 'American way' produced a 'civic expectation' that private initiative would serve communities through the development of non-governmental institutions. It was this that explained the emergence of numerous voluntary community organisations and national non-profit organisations and foundations.

Partnerships can be 'pluralistic' (such as in Baltimore where citizen groups were involved in urban renewal) or exclusivist (where government and the private sector may join in property development partnerships with minimal citizen involvement). In general, pluralist partnerships encourage the incorporation of various interests at the local level, but the patterns of incorporation differ from one partnership to another.

Public–Private Partnerships

The GTF report defined three basic areas of public–private coordination partnerships, philanthropy and voluntarism. There was overlapping between these areas while each had a distinctive role to play. In this context, public–private partnership was described as

> collaboration between business, non-profit and governmental organisations for the benefit of the partners and of the broader community. The idea behind partnership is simple: that through cooperation public and private sector organisations can, in many instances, serve their own needs and those of the community more effectively than by working alone. As such, public–private partnership grows out of the recognition by the public and private sectors of shared needs and common goals.
>
> (Commonwealth of Massachusetts, 1983, p.4)

Risks, resources and skills are shared in joint projects. Each partner is thus regarded as an active element and as one which may enhance the potential of the partnership by way of its contribution. In general this involves:

1 the active involvement of personnel from each of the participating organ-isations;
2 the clear identification of the partner organisations within projects and programmes; and
3 the presence of 'relatively direct' and identifiable benefits to the partners as well as to the community as a whole.

This definition excludes the conventional purchase of services or contrac-tual relationships which are essentially straightforward forms of business activity. Direct grant or subsidy relationships, where there is no element of personal collaboration, are also excluded in the GTF definition together with public–private partnerships where there is no private sector partici-pation (such as between two local authorities) and private initiatives such as company training programmes that do not involve outside organisations or public agencies.

Non-profit Initiatives

The GTF report's definition of 'philanthropy' covered the financial sup-port (through individual, foundation-based or corporate giving) of insti-tutions which offered programmes to strengthen the social fabric or improve the quality of life in a community. The philanthropic tradition defined by the report was that of mutual aid directed at the development of social and cultural issues and its direct reflection of individual and organisational concerns which lie beyond government. The continuation and expansion of corporate charitable giving was seen to be vital if non-profit organisations were to expand and survive.

The term 'philanthropy' may seem a little outdated in the 1990s. Since the publication of the GTF report there has been a great diversification of charitable and non-profit interventions in both the USA and Britain. It is thus appropriate to talk in terms of corporate and foundation support for non-profit organisations and initiatives. In this way, it is possible to identify different approaches to establishing relationships involving non-profit organisations.

There will inevitably be an overlap with the public–private partnerships described above. The essential distinguishing characteristic of partnership, as opposed to traditional philanthropy, relates to the diversity of relation-ships between those involved. Traditional 'philanthropic' activity usually involves a relationship in which fund providers act to back or support fund recipients. In the partnership situation, as initially defined here, there is more of a two-way and developmental approach to expanding the role of the partnership.

It is also possible for non-profit organisations to develop their activities and to work within public–private partnerships by adopting objectives which envisage the development of commercial-style operations within communities. In such cases, partnership is generally distinguished from traditional philanthropy and voluntarism by the elements of self-interest, organisational identification with wider public–private projects and cooperation with a variety of agencies in the planning, operation and oversight of programmes.

Volunteerism

The third type of activity identified in the GTF report is 'volunteerism' which involves the support of institutions that use volunteers or rely on volunteer management boards. In the USA and Britain this usually involves employees in companies giving their time on an unpaid basis to community organisations, charities or other types of non-profit organisation which themselves may be involved in local partnerships. Many companies have encouraged their employees at all grade levels to get involved in communities and to assist with the development of 'voluntary' organisations.

Partnership Models

Given the definitions of public–private partnership activities above, the GTF report classifies partnership arrangements in terms of their essential *purposes*. There are different forms and patterns of association within partnerships between the public and private participants and non-profit organisations. Partnerships can be between business and government, business and non-profit organisations, non-profit organisations and government, or between all three groups. Partnership can be 'loosely categorised' by purpose. First, there are deliberative or policy partnerships which are defined as those which, in the main, aim at the articulation of common problems and goals. Second, there are operational partnerships which are joint efforts in which such goals are actively pursued and where projects and programmes are implemented 'on the ground'. The types of partnership activity can be placed on a simple matrix showing actors and purposes (see Figure 9.1).

The matrix in Figure 9.1 should not be taken as providing a single or rigid format. It is indicative of only four possible combinations which 'can be tailored to meet an assortment of specific needs' (Commonwealth of Massachusetts, 1983, p.10). This applies equally if the matrix is used to describe American or British partnerships. The matrix is useful in the defining of different partnership contexts and stages of their development. The matrix can be expanded and enhanced to include innovative structural

	Policy	Operational
		Purpose
Business/Government	x	
Business/Non profit[1]		x
Non profit[1]/Government		x
Business/Non profit[1]/Government	x	x

Figure 9.1 Public–private partnership combinations

[1] including foundations
Source: Commonwealth of Massachusetts, 1983

arrangements, different mixes of policy and operations and local variants which may involve international organisations.

The purposes defined in the simple matrix should also be seen as providing a general description of the *main focus* of the partnerships. It may be that a business–government deliberative partnership is initially formed in a town or city to set goals for future action. However, at a later date, the partnership may start to concentrate more on operational activities and become involved with community organisations and non-profit agencies. Partnerships thus tend to change their organisational, managerial and financial activities as they develop. The GTF examples in the matrix shown thus relate to the essence of particular partnerships at a particular point in time.

The changing and expanding partnerships discussed below illustrate the complexity and the variety of political networks that exist. The examples of non-profit initiatives provided in Chapter 10 further indicate the rich variety of interactions and relationships in initiatives, many of which fall outside the matrix provided in Figure 9.1.

It is important that partnerships do change as they develop. Generally, partnerships are formed in order to overcome defined problems which exist at a particular point in time. However, it is not enough for public–private sector initiatives to remain in communities as 'one-off' problem-solving devices. The contribution which such initiatives make, or should make, is in the development of more sustainable structures which can assume a positive developmental role in communities and which harness resources that would not otherwise be mobilised. In this respect, there is a substantial amount of evidence from the USA that partnerships can provide tangible and lasting benefits to communities and can effect important new political alignments at the local level.

CREATING NEW STRUCTURES

Baltimore provides one example of how public–private partnership can produce tangible improvements for communities. It is important for both the private and public sectors that partnerships produce gains for communities which are perceived by those communities to be beneficial to them. The Baltimore example is in many ways a 'classic' in this respect as far as the private sector in Britain is concerned. It, and similar examples, have provided British companies with an indication of what can be achieved by mobilising resources and political support for initiatives which fall outside traditional political arrangements. Baltimore has often been seen as providing the living expression of a form of consensual politics which has defined 'community action' in terms of participation and cooperation.

Cowie (1985) examines the Baltimore experience for a British company audience. Baltimore stands as one example of a regeneration process which has 'naturally' developed an 'enterprise culture' and a balance between public and private interests. In its early days, the notion of local partnership was one which represented a clearly defined community orientation. In Baltimore, the initial stimulus behind the regeneration process came from the Citizens' Planning and Housing Association (CPHA) which was originally established in 1941. The association represented the coming together of a group of architects, planners and university professors, together with around 150 voluntary organisations of various kinds.

In the 1950s, the CPHA campaigned for the creation of an integrated development agency and was involved in the formation of the Greater Baltimore Council (GBC). The GBC consisted of the chief executives of the 100 largest corporations in the metropolitan Baltimore area. It was they, together with the CPHA, who seem to have effected a 'synergy' which produced the later dramatic improvements in the economic condition of the city and a feeling of pride in the new civic regeneration (HUD, 1979, pp.80–2).

The business community in Baltimore was in a good position to direct public attention towards the effects of urban decline. In doing this, company executives made the connection between decline and the longer term condition of the Baltimore economy and the fate of local businesses in the area. There was seen to be a direct link between the viability of local business and the level of activity in the local economy. There was also a relation between the ability to attract inward investment into the city and the general perceptions which abounded about the condition of the city. If a dynamic approach to regeneration could be engendered, then there could be a greater impulse towards growth and a reinvigoration of the economic base of the city. These were all considerations which have since been relevant in partnership initiatives in Britain and the USA.

The GBC produced a development plan that concentrated on the need to develop business activity in the run-down downtown business district. It involved the identification of areas that required extensive urban renewal, and included the assessment of the overall development potential of Baltimore's central area. A thirty-three acre area was designated for the first renewal initiatives. The cost was initially US$127 million which was to be partly raised by the issue of a public bond floated by the city for land acquisition. Around US$90 million was to come directly in the form of commercially-based investment in property and construction.

The early development, in what became the Charles Centre, was followed in the 1960s by a new partnership designed to regenerate the Inner Harbour area and to link it to the Charles Centre by extending the city's commercial district. This was a public sector led scheme which involved a substantial element of public residential development. There was a conscious attempt to encourage people in public housing to live in the regenerating central area to reverse the trend to develop subsidised housing in suburban districts.

There was also substantial private sector investment in the Inner Harbour project. Large corporations, such as IBM, constructed buildings in the area which now complement the Maryland Port Administration's World Trade Centre. The stimulus provided by the public sector was thus an important factor in encouraging further private investment in commercial and leisure-related schemes. The confidence generated by this investment therefore had a 'knock-on' effect in terms of the further development of the area through the commitment of additional private and public sector funds.

Diverse Approaches

The variety of partnerships and coalitional activity in the USA illustrates the adaptability of the partnership models to particular local needs and circumstances (Bingham and Blair, 1985). Each local variation is different in terms of finance, administration, structure, participation, objectives and so on. Partnerships can exist in the largest cities and smaller towns, as in the old textile city of Lowell in Massachusetts.

In Lowell, local political leaders and the city's banking community joined together in 1975 to generate public and private funds for urban regeneration following the dramatic decline in textiles. The Lowell Development and Financial Corporation (LDFC) was the result. It provided start-up capital for private ventures at heavily subsidised rates of interest and loans for small businesses. The Lowell initiative was implemented with the participation of local community organisations. The results have been notable, and stand as an example of how partnership can make a big impact on the confidence and economic prosperity of formally run-down towns and cities (CBI, 1988).

PARTNERSHIPS AND 'COLLECTIVISATION'

The American examples indicate the continued importance of the role of the public sector as well as the part played by business leadership (see Bach, 1990 and Page, 1990 for discussion of possible expanding public authority leadership in the 1990s). In Baltimore, public support came from both local and federal governments although it was the private sector which provided the initial stimulus to the Baltimore development process. The private sector was able to use public funding (such as that available through the Urban Development Action Grant) partly as the basis to lever further private funds into projects. It was the pattern seen in other American cities, and it was the approach that influenced government thinking in the UK in the 1980s and early 1990s. In this way, the public sector support in initiatives acted as a facilitator to 'pump-prime' extra private investment.

Pump-priming was encouraged by both the Thatcher and Major governments in areas where public money was used to create much of the vital infrastructure that was needed before the private sector would come forward with investment. For example, the government designated the London Docklands Development Corporation (LDDC) to coordinate the activities of the private and public sectors, and provided substantial incentives to developers and companies wishing to operate in the Docklands area. The designation of an enterprise zone within the jurisdiction of the development corporation was one example of how the government expected to attract private interest through the provision of what were effectively 'non-market' incentives in the Docklands (Batley, 1989).

Local authorities were often reluctant to cooperate with the UK urban development corporations, but the financial benefits were attractive (*Estates Gazette*, 1988). Pump-priming and other public incentives were regarded as a method of governmental intervention in the market which attempted to create more favourable financial conditions for the private sector. These conditions (for example, the granting of financial incentives and the relaxation of planning regulations in enterprise zones) provided companies with opportunities which may otherwise not have existed (Bruninvels and Rodrigues, 1989).

Within public–private partnership arrangements, such as in the Baltimore case, the combination of such *non-market incentives* as part of a *collective effort* by governments and the private sector pointed to the development of an urban regeneration strategy which modified the 'free market' and favoured for-profit companies. Such approaches involved the provision of financial incentives together with joint (or 'collective') action of a kind which enhances the interests of those involved in the partnership. Companies may cooperate within initiatives by sharing information and resources and obtaining assistance from local and national governments. They may do this to further profit-related objectives and/or to promote stable market conditions (where urban disorders have been a problem for example). In this, the 'collectivisation' of effort which takes place within the partnership potentially enhances the strategic market position of the companies involved. The companies may benefit by seeing land prices in the partnership jurisdiction increase and the asset values of their investments rise. This may be in addition to the broader economic benefits associated with the city wide regeneration which may accompany the implementation of the partnership strategy over time.

BUSINESS IN THE COMMUNITY

This was one motive behind the formation of Business in the Community (BITC) in the UK. In April 1980 representatives of business and government from Britain and the USA met at a conference in Sunningdale Park near London. The purpose was to discuss ways in which the private sector might help resolve some of the problems of the cities and how companies could benefit from corporate involvement in socially beneficial initiatives (BITC, undated, p.12).

A working party was established, serviced by the British Department of the Environment (DoE), and chaired by Sir Alistair Pilkington, the prominent industrialist. The working party concluded that collaboration between government and industry was of vital importance to the success of future initiatives. This would include the active participation of voluntary organisations and the trade unions (Pilkington, 1984). The approach set the scene for the promotion of a strategy for integrating the efforts of diverse interests around a common set of goals designed to restore confidence to the cities.

The establishment of Business in the Community, following from this initiative, promised to provide a framework which would stimulate the interest and encourage the support of companies for direct involvement in communities. Initially, BITC was supported by nine companies and a number of central government departments. Three years later, BITC had around seventy affiliate companies, and in October 1984 its activities were amalgamated with those of the Special Programmes Unit of the Confederation of British Industry. By 1988, BITC reported that it had 331 members which included 291 separate companies plus various voluntary and public sector organisations. It was reported that almost two-thirds of *The Times* top 100 companies were affiliated to BITC; many with direct links to USA companies or subsidiaries operating corporate social responsibility programmes.

In Britain, the 1981 urban riots added a new sense of urgency to the BITC initiative. The scenes of devastation in major towns and cities following the riots seemed to underline the urgency of the tasks facing decision-makers. The American experience became even more relevant to the British. The USA had seen the affects of riots in the 1960s and had responded with a strong private sector orientation in the cities. Could not the same be achieved in Britain?

Even before the riots, BITC had established links with Americans involved with the promotion of corporate social responsibility. By the mid-1980s BITC was gaining from these contacts in terms of how it perceived its own rather ill-defined objectives and how it saw itself developing in the future. There was a move away from its early concentration upon backing for business advice agencies (through the enterprise agencies), towards a broader conception of the role of company community activities (BITC, 1986). BITC came to regard its mission as attempting to effect a change in the 'culture' that affected change in the cities. It was, in short, a recognition of the kind of self-sufficiency philosophy that was being developed by the Thatcher government, although the BITC view was never cast in party political terms.

The aim was to bring about a change in attitudes in the cities and to develop new frameworks that would themselves assist with the development of community notions of enterprise and initiative (BITC, 1987a, 1987b, 1988). The overcoming of the kind of apathy mentioned by Cowie was an implicit objective. The 'enterprise culture', while not being explicitly promoted by BITC, was evident as a practical operational concept in virtually all of the organisation's partnership work. The major constraints placed upon the activities of the organisation seemed to relate to its limited resources, its difficulty in maintaining momentum in some areas of its activities and the problems faced in encouraging community activists effectively to coordinate their activities with those of BITC.

The CBI Task Force

The conceptualisation of the leadership role of business in communities was carried a stage further in Britain by the Confederation of British Industry (CBI). At its 1987 national conference, the CBI decided to establish a task force which would 'identify what further steps business should be taking to assist in the process of urban regeneration' (CBI, 1988, p.7). In 1988, the first report of the task force was published entitled *Initiatives Beyond Charity*. The report clarified the CBI's thinking on the role of business leadership in four main areas.

First, business had to provide the leadership in urban renewal to reverse what the report described as the 'cycle of economic and social decline' in Britain's cities (CBI, 1988, p.9). The quality of business leadership in any city would affect the degree of confidence within the city both within communities and amongst likely outside investors. The stress was upon the need to develop confidence, since it was that factor which was so important in creating the right environment for investment and the generation of economic activity. The building of confidence required the 'full commitment of leaders from many different quarters behind a common vision; but business leaders had a key role in providing necessary enterprise, energy and credibility to rebuild confidence amongst the business sector at large' (CBI, 1988, p.9).

Second, the report saw urban decay in Britain as a problem whose solution lay 'beyond charity'. The CBI was concerned that the process of urban decline could rapidly develop once started, and that an 'underclass' of people who were virtually excluded from the rest of the economy could easily emerge. This was seen to be a characteristic of urban decline in London and other major cities, and was regarded as a high cost problem in terms of its social and economic consequences. For business, the economic implications were that such decline tended to reduce the spending power of communities, led to training and recruitment problems in relation to the unemployed (particularly with ethnic minorities) and wasted development opportunities for companies in areas affected.

Third, the report stated that there was a need for *a process* of renewal that would lead to sound economic development (CBI, 1988, p.10). The envisaged process was one that would be applicable in a variety of situations where there was an evident need for urban renewal and where there was a potential for mobilising local resources and community effort. This required the creation of visible 'flagship projects', the consolidation of 'good practice' and 'early wins' in initiatives. The emphasis was on the development of a 'self-sustaining momentum' where business leadership teams would facilitate the integration of the many individual effects which would come together in community projects.

Market-Led Renewal

The report was clear about the basis for this strategy. It was the development of market-led initiatives which were to be of importance. The new structures that developed within communities would be dependent upon the creation of an economic infrastructure which would be essentially a profit-responsive one. This required that the main criteria for evaluating the success or otherwise of urban renewal was one which related to the commercial viability of public–private partnerships. This would be combined with a strong element of enlightened self-interest for companies reminiscent of the approaches adopted in various public–private partnerships in the USA. In addition, the implication of the CBI report was that both public–private sector partnerships and non-profit initiatives could be seen as enhancing the interests of companies in this setting. This was clearly evident where the report identified the 'spin-offs' from urban renewal partnerships:

a) the promotion of land and property development opportunities;
b) the creation of improved job opportunities and the development of a more skilled workforce through the expansion of education and training;
c) the greater involvement of communities in urban regeneration, crime prevention, educational compacts and charitable activities;
d) the creation of public/private sector partnerships which would be closely integrated with the business community's efforts and which would represent the development of a common approach to the problems of the inner-cities.

(CBI, 1988, pp.10–11)

Business Leadership in Action

The report advocated business leadership based on the above objectives. It called for determined action and stressed that the motive force behind future policy initiatives in urban renewal should be the private sector. Business was seen to have the energy and drive to achieve positive results particularly if there was 'an environment of business confidence' (CBI, 1988, p.13). Indeed, it was *only* where such a condition could be met that *public investment* would become productive and conducive to the generation of improved economic activity.

In addition, there was a critical political aspect involved with this approach. The report stated that it was 'simply unrealistic' to expect local politicians to be able to assume this kind of leadership role on their own. There was no tradition in Britain of the American-style executive mayor who could play an important role as 'facilitator' in bringing about substantial improvements in the cities. The report was aware of the limitations of British local government and worried about the implications of the high rates of turnover among elected members of local authorities and their senior full-time officers. Baltimore's celebrated renewal programme, by contrast, was led by a mayor who held office for over a decade and later became the governor of the state.

The CBI report therefore suggested that it was a combination of structural and political problems in British local government that contributed to the difficulties in regenerating the inner-cities. This implied a significant criticism of the British approach to urban renewal and implicitly suggested that American local government was more effective with important decisions being taken by a relatively small number of expert professional decision-makers. The report therefore favoured a more clearly identifiable leadership element within cities that would operate with extensive discretion over important issues of public policy. However, unlike many partnership initiatives in the USA, there was no indication in the CBI report as to precisely how the participants in British partnerships would be accountable to elected representatives or how business leadership groups would relate to existing local government institutions.

The CBI was critical of local politicians who could not provide what the report described as 'the visionary leadership required' in partnerships. This criticism also extended to central government bureaucrats and even to ministers. Worthwhile initiatives would have to involve local government and voluntary agencies, but it was evident that the 'vision' was one which the CBI saw as coming from the private sector (CBI, 1988, p.13; Calderdale Partnership, 1989).

The implications of the CBI view are profound, especially if the report is taken at its face value and is not to be seen as merely staking claim to a bargaining position in relation to local and central government. There was emphasis in the report on criticism of the past records of local authorities, whereas in the USA this distancing of companies from local governments seems to be less overtly stated in public. In the USA, local and state governments have not been associated with the kind of leftist political strategies that many city administrations in Britain have experienced. This may be one reason for the CBI's apparent coolness towards local politicians and bureaucrats. Also, in the USA, city and state governments have a long record of providing conditions that are conducive to the development of free enterprise and the encouragement of a favourable business environment. That has not always been the case with Labour controlled urban local authorities in Britain.

Nevertheless, the CBI called for a change. Local government had the opportunity to join with the private sector in developing the new 'shared vision' of the future and in developing mechanisms, outside traditional local government structures, to facilitate this. New organisational arrangements would be accompanied by the development of new and innovative managerial structures and financial packages which would involve the efforts of small teams of very senior business people 'of stature' within communities. Therefore, at the top of the business leadership tier there would be a group of senior personnel who would have an important stake in the locality and who would have contacts with others of similar status through a national network operating between leaders in various British towns and cities. Local teams should thus consist of no more than ten people and should seek to co-opt the chief executive of the relevant local authority with the support of the governing party.

A Shared Vision

The perspective depended upon the creation of a consensus which would provide the conditions for dynamic change. It was this condition that appeared to be characteristic of the American partnerships, although there was no reference in the CBI report to those instances in the USA where partnerships had created tensions between the participants. In such cases there was often a conflict between the aspirations of community organisations and the objectives of large corporations, particularly where city planning or environmental issues were at stake. Sometimes it was groups external to the partnerships in the USA which were critical of the strategies being adopted by those who had the equivalent of the CBI's 'shared vision'.

In Britain, many Labour controlled local authorities were expected to be suspicious of business leadership teams that were to operate outside traditional local government. However, in spite of this there was some substance to the CBI's hope that consensus could prevail. There was the experience of BITC in obtaining the active support of the trade unions at the national and local levels. There was also evidence which pointed to a supportive attitude for public–private partnerships from even quite militant left-wing Labour councils. Labour politicians were usually very enthusiastic about the prospects for economic and community development within their jurisdictions, and so the partnership approach could be 'sold' relatively easily to those activists who were determined to enhance the quality of life in distressed city districts.

Action programmes were to be the binding force behind the partnerships proposed by the CBI. It was through the development of action programmes that the task force saw the partnerships being able to engender a common community purpose and where there would be a convergence of objectives shared by the parties involved in the partnership. One aspect of this related to the development of commonly held views about how particular cities would develop in future years and about the kind of place cities ought to be. This would involve the partnerships in generating employment, increasing community spending power, improving the environment and increasing civic pride. In this, there was, therefore, an appeal to a wide variety of community interests, ranging from trade unions through to the local green lobby.

Difficulties with the CBI approach were, however, related to how this could work in practice over a longer period of time. The evident strains and tensions that appear in partnerships could threaten the essentially localist coalitions envisaged by the task force. Political conflicts could emerge especially where there were disputes over resources or constraints on expenditure (Moore and Richardson, 1989). There could also be problems associated with conflicting loyalties where decisions had to be taken which would affect the competitive situation between towns and cities seeking inward investment or other kinds of economic input into local economies.

Modern corporations do not operate in single localities. The majority of large companies now view their operations in national and international terms. During periods of economic growth it may be relatively straightforward to adopt a local orientation, but when times are less favourable hard decisions may need to be taken which may adversely affect particular communities. What happens to the local 'vision' if a company is forced to choose between two towns when there is the prospect of plant closures? Even if the company is concerned to limit the adverse impact of a closure by way of an intentionally socially responsible policy on redundancies, there will tend to be a disruption to the smooth operation of the local partnership process. By definition, plant closures act against the basic objectives inherent in the CBI action programmes and reduce the capacity of communities to generate sustained economic activity in the short term.

Action at the Local Level

The next step after the publication of *Initiatives Beyond Charity* was to implement the report's main findings. This involved the development of strong links with local political leaders and government officials, and joint action with BITC and a range of other organisations. This was achieved through the formation of Business in the Cities which was a direct result of the CBI report. The new forum met for the first time in December 1988, consisting of representatives from the CBI, BITC and the Phoenix Initiative (Business in the Cities, 1988a, 1988b, 1990).

The forum was concerned that there should be no duplication of effort between the organisations involved at the national and local levels. The aim was to establish business leadership teams in various parts of Britain which would draw upon the expertise of the founding agencies. These were initially located in Bristol, Newcastle, north east London, north west England, Sunderland and Teesside with a so-called 'watching brief' over the Birmingham Business Initiative (The Newcastle Initiative, 1990). There was also to be involvement with leadership groups in Blackburn, Calderdale and Sheffield.

The local teams were often simply those with whom BITC had already been involved. In this way, the support of BITC was a key factor in enabling Business in the Cities to begin to coordinate a national effort and effectively to create a further broadly based national network of agencies concerned with urban regeneration. This seemed to compliment BITC's earlier (1987) formation of eight national target teams intended to promote flag ship projects and demonstrate the creative role that business could play in urban renewal (BITC, 1986 p.3). The target teams covered areas including priority hiring, education, finance for enterprise, enterprise development, land purchasing, the voluntary sector and marketing. Business action teams were established to bring together senior members of the business community with other local parties, and it was around some of these action teams that the later business leadership teams were established.

Redefining Corporate Social Responsibility

The prospect was one that envisaged the development of new local and national networks of organisations concerned with the future development of Britain's cities. This added to the already complex urban renewal process existing in Britain. As the Audit Commission (1989) pointed out, urban renewal involved a process that was characterised by the fragmentation of the policy process, and a situation in which many local authorities believed that their role was being undervalued by central government. Many councils were of the opinion that they were being marginalised as far as their place in urban regeneration was concerned.

The Audit Commission's response to this was, not surprisingly, to call for an essentially market-led approach in which local authorities would coordinate their activities with central government and the private sector. The political climate at that time was still heavily on the side of the market and suspicious of significant infusions of money into public sector initiatives. The Audit Commission also saw a role for the development of partnerships through the urban development corporations (UDCs), despite the lingering tensions between some local authorities and the UDCs (Cowan, 1987). However, the question arises as to just how far the dependence upon the market in public–private partnerships will lead in the longer term to a redefinition of the concept of corporate social responsibility.

Profit-generating developments are regarded as being beneficial to communities particularly if they improve the environment or create social facilities in areas (Business in the Environment, 1990). There are numerous examples of developments where corporations have been seen to provide landscaped areas around their headquarters buildings, where leisure facilities form an integral part of commercial and business parks and where the attractive combination of leisure and retailing appear to enhance the quality of many formally run-down inner-urban areas. This appears to be the market acting as the socially responsible agent, almost by definition. It is part of the allure of the popular market philosophy that was discussed in Chapter 1.

This is an aspect of social responsibility which links to various policy areas relating to how companies can present themselves as environmentally friendly; as being caring for the community; as responsive to local aspirations and demands, and aware of the need to modify planning strategies to take account of the needs of ordinary people. It is the redefinition of profit as one of the most socially responsible ways of ensuring the future of the cities and the well-being of their inhabitants.

CONCLUSIONS: CONSTRAINTS AND LOCAL GOVERNMENT AUTONOMY

In the USA, the range of state and local government incentives available to companies operating within various kinds of public–private partnerships is conducive to the promotion of economic growth through urban renewal (for example, City of New York, 1983; New York Chamber of Commerce, 1983). There are also a range of financial and legally defined development 'models' which give American companies and non-profit organisations considerable flexibility when it comes to developing projects and programmes specifically designed to meet particular local requirements. For example, the Americans enjoy important tax advantages which can be maximised by establishing any one of a range of non-profit organisations.

The 'non-profit corporation' as such does not exist in the UK, despite attempts by some organisations in Britain to create operating structures similar to American non-profit organisations by working within British tax laws (for example, under the regulations governing charities and the formation of charities with corporate status). In the USA, there are often opportunities for business-led public–private partnerships and non-profit initiatives to take advantage of special below market rate of interest loans to developers both from federal programmes and development corporations. The opportunities for utilising below rate loans in Britain are severely constrained, and generally only obtainable by large companies through the banks or other financial institutions (and then within strictly defined limitations).

The non-profit sector in the USA enjoys considerable flexibility (see Chapter 10). This is especially so with respect to the range of funding and investment opportunities that are legally available. It can be argued that in this way the American non-profit organisations are able to operate within a more liberal market setting which encourages initiative, innovation and capitalist entrepreneurialism which would be the envy of many non-profit organisations in Britain (Jacobs, 1989b). This places American non-profit organisations in a position where they can develop their 'commercial' activities and evolve a more clearly defined market orientation.

Another factor when looking at the funding of public–private partnerships is the position of local government authorities. British local authorities have been severely restricted with regard to the discretion which they have over the funding of local programmes. Local authorities have been hit by controls on their capital spending, revenue spending cuts and a host of regulations affecting their commercial and economic development activities. These have constrained British local government from making substantial 'independent' financial interventions in urban renewal, or where attempts have been made, they have faced difficulties within the confines of a severely constrained financial regime.

Where councils in the UK have embarked upon ambitious programmes there have been repercussions politically as the burden of commitments has fallen directly on local people. For example, Sheffield City Council's decision to host the 1991 World Student Games involved the local authority in a major series of construction projects for the games. Central government gave little encouragement to the Labour controlled council. Funding was sought from a reluctant private sector, while a substantial amount of the funding had to be found from the public sector thus placing a burden on Sheffield's Community Charge payers.

Economic Factors

The development of partnerships and the 'shared vision' of corporate involvement pose a challenge to local governments in Britain and the USA. Leadership teams along the lines proposed in the UK, and the powerful growth coalitions in the USA, operate outside the established arrangements of local government. They involve a variety of interests, yet they are not directly accountable to the wider electorate.

The development of public–private partnerships has taken place at a time when economic changes have affected the location decisions of major companies and when property development companies have assumed an important role in city regeneration strategies. The major participants in development projects may not have their headquarters in the cities within which they are investing. International companies operate in property development and affect decisions about projects in particular cities. Chapter 2 suggested that local governments are therefore heavily influenced by the market decisions of such companies. This affects the locus of real decision-making in the cities; either shifting influence over strategic planning and economic policies in the direction of interests external to the city or enhancing the influence of non-local government leadership groups.

As suggested in Chapter 2, local governments develop their own interests. They are not simply powerless in the face of these challenges, although their influence can be significantly reduced by market pressures. Local authorities have often taken an important role in local partnerships and their financial and facilitating role is often crucial in getting partnerships started. Local policy communities vary in terms of their composition and the ways in which they operationalise their policy objectives. As a result, the role of local government differs from place to place and local 'models' each have their own specific characteristics. Such variations, however, should always be examined by reference to the wider context of change.

NOTE

Numerous organisations were contacted and representatives interviewed (including council members and public officials) in connection with the research for this and other related chapters. Those which were of particular assistance are included below.

The United States

The Urban Institute; The Greater Hartford Chamber of Commerce and Hartford City Council; the Hartford Seminary Foundation; Boston City Council; ITT Corporation; Time Incorporated; The Goodyear Tire and Rubber Company; General Motors Corporation; US Department of Housing and Urban Development; National Association of Manufacturers; Independent Sector; the Federal Reserve Bank of Boston.

Britain

Business in the Community; Local Initiative Support (UK); British Petroleum; The Phoenix Initiative; Centre for Employment Initiatives; Wolverhampton Borough Council; Stoke-on-Trent City Council and Community Partnership; Staffordshire County Council; the Department of the Environment; Merseyside Task Force; Sheffield City Council; Sheffield Development Corporation; British-American Tobacco; Charities Aid Foundation; Training Agency and Peat; Marwick Mitchell.

10

INITIATIVES BEYOND CHARITY

The 'non-profit sector' refers to non-governmental organisations which do not form part of the corporate private sector. These are commonly 'voluntary' or charitable in nature. However, the problems of defining such a sector are many. The terminology can be confusing. The term 'non-profit' here describes organisations that are distinct from the corporate private sector and which operate as charities, foundations and non-charitable voluntary organisations. The use of the more restrictive term 'charity' largely describes organisations which are legally constituted for charitable purposes (Ware, 1989, p.3).

Non-profit organisations play an important part in the promotion of urban initiatives. They act as points of contact between the corporate private sector, educational institutions, communities and governments at the national and local levels (TAFT, 1987; Hoy and Bernstein, 1981; Council for Financial Aid to Education, 1986 for example). They facilitate the implementation of private sector and governmental policies in urban renewal and in the delivery of contracted services. In this, non-profit organisations are

1 increasingly regarded by governments and the corporate private sector as vehicles for promoting the 'empowerment' of local communities;
2 aware of the commercial implications of market approaches to supplying local services and generating wealth in communities;
3 concerned with the monitoring and evaluation of their activities to improve managerial performance and competitiveness in a market environment where governments privatise and contract-out services;
4 developing links through 'umbrella' organisations to enhance resources and benefit from closer relationships with fund providing organisations.

In the UK, non-profit organisations have assisted the initiatives pursued by government which have stressed the importance of self-help and enterprise. In the USA, 'empowerment' has been one of President Bush's domestic policy concerns; promising the exercise of community influence

216

over a wide range of local services and programmes ranging from tenant management and schools choice to the environment, health and social service provision. Empowerment has been a vague and often misused concept. However, with the establishment of a Republican party task force on empowerment (chaired by Housing Secretary Jack Kemp), the concept was adopted as part of the social vision of the free market.

Non-profit organisations were praised by governments for promoting voluntary effort, local involvement and community regeneration. They were regarded by free marketeers as being conducive to the development of the new corporate culture and the 'shared vision' of business-led urban renewal. They could thus play an important part in creating the conditions for the growth of the wider private sector both domestically and internationally. President Bush's 1991 state of the union address contained a reference to the 'thousand points of light' which were part of the American spirit of philanthropy and which should be at the centre of economic renewal.

THE 'INDEPENDENT SECTOR'

In the USA, the 'independent sector' has been described as consisting of those non-profit organisations defined by the 501(c)(3) and 501(c)(4) categories under the federal tax code for tax-exempt organisations (Independent Sector, 1989). The definition covers a variety of such tax-exempt organisations in addition to other non-profit organisations not covered under these tax categories.

Non-profit organisations have found it useful to co-operate. In 1980, the Coalition of National Voluntary Organisations (COVO) and the National Council on Philanthropy (NCOP) created a formal organisation named Independent Sector (IS). This aimed to preserve and enhance the American tradition of giving, volunteering and non-profit initiative characteristic of 501(c) organisations (IS, 1990). The IS had 700 member organisations at the end of 1990.

US Non-profit Organisations

American non-profit organisations obtain a substantial portion of their money from private sources. This is largely in the form of contributions of money, 'in kind' contributions and the sale of products and services. An illustration of some of the funding sources of US foundations and the characteristics of four types of foundation are given in Table 10.1.

Estimating the number of institutions in the US 'independent sector' is difficult. It includes the tax-exempt organisations as well as religious institutions which are not required to report to the federal tax authorities. There are also several thousand organisations defined as public charities

Table 10.1 General characteristics of four types of foundation, USA

Foundation type	Description	Source of funds	Decision-making body	Grant-making activity
Independent foundation	An independent grant-making organisation established to aid social, educational, religious, or other charitable activities.	Endowment generally derived from a single source such as an individual, a family, or a group of individuals. Contributions to endowment limited as to tax deductability.	Decisions may be made by donor or members of donor's family; by an independent board of directors or trustees; or by a bank or trust officer acting on donor's behalf.	Broad discretionary giving allowed, but may have specific guidelines and give only in a few specific fields. About 70 per cent limit giving to local area.
Company-sponsored foundation	An independent grant-making organisation with close ties to the corporation providing funds.	Endowment and annual contributions from a profit-making corporation. May maintain small endowment and pay out most of contributions received annually in grants, or may maintain endowment to cover contributions in years when corporate profits are down.	Decisions made by board of directors often composed of corporate officials, but which may include individuals with no corporate affiliation. Decisions may also be made by local company officials.	Giving tends to be in fields related to corporate activities or in companies where corporation operates. Usually give more grants but in smaller $ amounts than independent foundations.
Operating Foundation	An organisation which uses its resources to conduct research or provide a direct service.	Endowment usually provided from a single source, but eligible for maximum tax-deductable contributions from public.	Decisions generally made by independent board of directors.	Makes few, if any, grants. Grants generally related directly to the foundation's programme.
Community foundation	Most often a publicly supported organisation which makes grants for social, educational, religious, or other charitable purposes in a specific community region.	Contributions received from many donors. Usually eligible for maximum tax-deductable contributions from public.	Decisions made by a board of directors representing the diversity of the community.	Grants generally limited to charitable organisations in local community.

Source: The Foundation Directory, 8th edn, published by The Foundation Center, 79 Fifth Avenue, New York, NY 10003–3050, 1981 and Jacobs (1987)

with annual revenues of less than US$500 which are not required to apply for tax-exempt status. There are also the social responsibility programmes of corporations which file different tax forms (and which do not always figure in statistics about the sector) and non-profit organisations that file omnibus tax forms covering many separate organisations (such as the Boy Scouts).

Of the estimated total of 1.3 million non-profit organisations in 1987, the US Department of the Treasury reported approximately 939,000 on its master file. Of the independent non-profit organisations covered in the 501 series, the Internal Revenue Service (IRS) reported that 422,103 in 1987 had 501 (c)(3) tax-exempt status. These were engaged in education, research, science, religion, philanthropy and other charitable activities. Other tax-exempt organisations serving primarily charitable purposes included the 138,485 organisations in the 501(c)(4) category covering mainly civic leagues and social welfare organisations. The 501(c)(3) organisations were legally entitled to receive tax-deductible contributions for both individuals and corporations, whereas 501(c)(4) organisations were tax-exempt but not generally in receipt of tax-deductible contributions (IS, 1989, p.14). In 1987, the independent non-profit sector's share of national income was 5.8 per cent or US$247.1 billion (IS, 1989, p.15).

UK Non-profit Organisations

The 1989 edition of the Charities Aid Foundation (CAF) publication, *Charity Trends*, reported over 165, 000 charities registered with the British Charity Commission. This excluded voluntary organisations which for various reasons did not register for charitable status but which were part of the broader 'voluntary sector'. The range of non-profit activities is indicated in Table 10.2 which shows the main non-government bodies involved in communities in Britain.

The CAF survey looked at the accounts of organisations with a voluntary income of over £150, 000. The 'charity' section of the CAF statistics was based on a survey covering 400 organisations. The survey obtained a 93 per cent response rate from the top 200 UK charities and a 91 per cent response from the second 200. Details about the accounts of some charities were obtained from CAF records, while others came from the files of the Charity Commissioners. Charities were asked if they were able to give a figure for money received through payroll giving. Thirty-nine received a total of £827, 000 which was regarded as an underestimate as some charities were unable to provide information.

Table 10.2 Main non-government bodies involved in community and related activities in Britain

Organisation type	Main support	Nos in existence	Activities
Non-charitable 'voluntary' organisations (ie. non-governmental bodies not registered as charities)	Funding and 'in kind' support from companies, national and local government, quasi-public bodies, charities and individuals	Over 160,000 voluntary and charitable bodies	Political, artistic, recreational environmental and social activities, including at community level
National and local charitable organisations (registered as charities)	Funding and 'in kind' support from companies, national and local government, quasi-public bodies and individuals		Wide range of social, religious and other activities which to various degrees assist private and public urban development projects
Councils for Voluntary Service	As charities and voluntary organisations	Approx. 200 in major towns and cities, etc.	Coordinate activities of voluntary and charitable organisations in localities
Private and public companies (especially those with public affairs departments)	Corporate support for public affairs and 'corporate' social responsibility	No reliable estimate of number of company involvements	Support for enterprise agencies, Business in the Community and local organisations including charities
Small businesses, cooperatives, etc. and City Farms	Funding and aid from companies, enterprise agencies, national and local government	No reliable survey of global involvement, about 35 City Farms	Participation in partnerships at local level between the public and private sectors and with charities etc.
Business in the Community	Major companies, banks and city institutions plus prominent individuals	About 250 affiliated companies	Wide support for and initiation of projects, the enterprise agencies, etc.

Source: Jacobs (1989)

The Role of Non-profit Organisations

These statistics, together with additional research, indicate that non-profit organisations are of considerable importance in economic terms (CAF, 1989; IS, 1988; 1989). Despite conflicts between the interests of non-profit organisations and others, charitable and voluntary organisations are an elemental part of the market economy (King, 1989, p.36). They help the corporate private sector to integrate social responsibility into their 'shared vision' through coalitions with community organisations and national voluntary groups. In many cases, non-profit organisations have strengthened local partnerships and facilitated the cooperation of groups with local and national governments.

In Britain, 'traditional' philanthropy expanded during the industrial revolution and was embodied in late-Victorian conceptions of self-help. Self-help was advocated as a solution to the problems of the poor, while the market provided the dynamic for overall economic growth. There emerged a desire to cater for neglected social provisions, either through state support, or by charitable involvement in matters of concern to the community. Charities 'filled gaps' where social provisions were inadequate, although they were often regarded as being of marginal significance (Jacobs, 1989b, p.84). They offered educational, social and welfare services that were important in the development of notions of socially supportive service provision.

Charles Booth, the wealthy Liverpool shipowner exemplified the Victorian spirit of philanthropy combined with a social conscience. Booth was confident that capitalism could benefit from the expansion of self-help and individual effort in cases where there was a moral justification for providing social supports to the poor. The idea of self-reliance was reflected in the thinking of many charities and supported by private companies as an early form of corporate interest in urban communities (Jacobs, 1989b).

'Classlessness'

During the nineteenth century it was sometimes suggested that the middle class shared an interest with the urban poor in bringing about improvements in health and sanitation and social conditions. If the cities were unhealthy and dirty, that was an issue which affected all classes. The middle class was concerned about the diseases caused by the insanitary conditions in the cities, where they had to conduct their business whether they liked the prospect or not. Charity was thus linked to this basic desire to survive and ensure better health in a polluted and unhealthy urban environment (Engels, 1962, 1979).

It was in this sense that charitable and non-profit provisions lent themselves to being seen as of benefit to 'the community' as a whole. A modern version of this tends to be propounded by community activists when there is a coming together of corporate interests with labour and voluntary groups. The *déclassé* quality of initiatives is also evident where 'world' or 'international' interests can be identified with 'the community', this being implicit in the UK Business in the Environment initiative (BITC, 1990a) and other 'green' campaigns. This should not imply that all charities and voluntary groups see themselves in *déclassé* terms, but there is a tendency for them to do so where 'community' and 'partnership' are used to describe commonly held objectives.

This is an attractive way of viewing joint activity. However, the simple class-neutral view conceals the essence of most partnerships. As seen in Chapter 9, partnerships consist of varied interests where the private sector can 'cartelise' the relationship with those involved. Various interests may be represented, but their full independent expression is often constrained. Groups tend to modify their demands to achieve the compromise necessary to consolidate the partnership (Jacobs, 1989b).

Non-profit organisations have different relationships with companies and governments. Partnership networks can be formal or informal. Some groups are directly linked to companies through sponsorship. Others may be independent but linked to companies or public bodies by contractual agreements. Other non-profit organisations may be associated with trade unions, churches and so on. The nature of partnerships involving non-profit organisations is thus diverse and determined by the local arrangements between groups.

GOVERNMENTS AND NON-PROFIT ORGANISATIONS

Organisations that do not make profits are sometimes viewed as suspect in market economies particularly when 'non-profit activity' is associated with mainly social provisions. Despite this, non-profit organisations can assist companies in various ways and have played a part in facilitating the transfer of social programmes away from government. This is consistent with conservative calls to limit the role of government in the provision of services (Ruckelshaus, 1982, p.30).

The expansion of non-profit organisations has also been related to the growth of a corporate social responsibility in the 1980s. Non-profit organisations were frequently mentioned in arguments favouring the enhancement of the role of the private sector in community affairs (Kotz, 1982; Atwater and Brewster, 1982; Verity, 1982). Thatcher's urban policy supported the growth of the activities of non-profit organisations and the Bush administration adopted a generally supportive attitude towards programmes which benefited them. In both the USA and UK, government funding was an important source of revenue for non-profit organisations particularly where they were involved in the running of services in conjunction with public sector agencies.

In the USA and Britain the activities of non-profit organisations therefore came to rely upon the support of government. This often produced a dependency relationship between the public sector and voluntary and charitable organisations which implied both costs and benefits for the non-profits (Jacobs, 1989b). One cost was that government support made them vulnerable to changes in political attitudes and funding commitments.

Government and US Non-profit Organisations

Salamon and Abramson (1990) show that the Bush administration continued to effect modest increases in federal spending in the US budget functions of special concern to private non-profit organisations. For example, after excluding Medicare and Medicade, there was a substantial growth in federal health programmes which benefited non-profit organisations under Bush. By fiscal year (FY) 1991, spending in this area was estimated at US$2 billion above the level projected in FY 1990.

The Reagan years had left a legacy of cutbacks in areas such as education, social services, employment and training, community development and international affairs. The consequences of the cuts were felt into the early 1990s despite the prospect of a slight easing in the funding situation under Bush. From FY 1982 to FY 1990, the inflation-adjusted value of federal spending in the six budget functions of greatest concern to non-profit organisations, exclusive of Medicare and Medicaid, declined a cumulative total of US$120.1 billion compared to what would have been spent had FY 1980 spending levels been maintained. The increases proposed in Bush's budget for FY 1991 therefore still proposed overall spending in the six budget areas exclusive of Medicare and Medicade of US$3.9 billion below FY 1980 levels (Salamon and Abramson, 1990, p.i)

The changes in government support for non-profit organisations had an impact on the revenues which they generated. Because of the cuts in the 1980s, the relative revenue situation in 1991 was projected to be 15 per cent under 1980 levels. For the FY 1982–90 period as a whole, non-profit organisations lost US$32.8 billion in federal support that would have been available had FY 1980 spending levels been maintained (Salamon and Abramson, 1990). Private giving was generally expected to offset the overall reductions in federal spending in fields of interest to non-profit organisations. Between FY 1982 and FY 1988, private giving offset only 41 per cent of the *cumulative* reductions in federal spending in these fields:

> Private giving has, however, managed to catch up with the direct federal revenue losses private non-profit organisations experienced in these fields. For the FY 1982–88 period as a whole, private giving

grew by forty-two billion US Dollars, compared to a 29.5 billion US Dollars shortfall in federal support. This is because Congress held the line on further cuts while private giving continued to grow. At the same time, there is little evidence that the growth of giving has been sufficient to cope with the growing needs left behind by the overall federal cuts and changing social and economic problems.

(Salamon and Abramson, 1990, p.i)

Table 10.3 shows the changes in federal spending in budget functions of concern to non-profit organisations. The situation described by Salamon and Abramson brought little relief to the non-profit sector as far as spending constraints at the federal level were concerned. This was despite Congressional concern, in early 1990, about the 'peace dividend' which was foreseen as the outcome of the diversion of funds from defence to other areas. However, the 'peace dividend' seemed destined to be used on federal deficit reduction rather than social provisions. The pattern of proposed spending alterations in the FY 1991 budget was also complicated by the fact that some programmes faced the prospect of substantial reductions, as in the case of community and regional development.

Table 10.3 Changes in federal spending in budget functions of interest to non-profit organisations, FY 1980–90, by year, after adjusting for inflation (in billions of constant FY 1980 dollars)

Fiscal year	Six budget functions[1]		Excluding Medicare/Medicaid		
	Outlays ($)	Change from FY 1980 level ($)	Outlays ($)	Change from FY 1980 Amount ($)	%
1980	200.5		151.5		
1982	195.0	− 5.5	140.1	− 11.4	− 7.5
1983	200.7	+ 0.2	144.8	− 6.7	− 4.4
1984	192.9	− 7.6	135.6	− 15.9	−10.5
1985	197.0	− 3.5	135.5	− 16.0	−10.6
1986	196.0	− 4.5	134.5	− 17.0	−11.2
1987	194.8	− 5.7	132.6	− 18.9	−12.5
1988	199.3	− 1.2	135.8	− 15.7	−10.4
1989	204.4	+ 3.9	138.9	− 12.6	− 8.3
1990(E)[2]	214.9	+14.4	145.6	− 5.9	− 3.9
Total FY 1982–90(E)	1,795.0	− 9.5	1,243,4	−120.1	− 8.8

Source: Salamon and Abramson (1990) based on US Government statistics
Note: [1] These budget functions include international affairs (function 150); community and regional development (450); education, training, employment and social services (500); health (550); Medicare (570); and income assistance (600). These functions account for approximately 30 per cent of all federal spending.
[2] (E) = estimated

Government and UK Non-profit Organisations

The government funding situation in the UK during the 1980s was one in which there was a gradual increase in government support for non-profit organisations as programmes were developed to accommodate organisations which were expected to take over activities previously handled by government. During the 1980s, the measured expansion of local and national government funding to charities was frequently accompanied by public and private sector support for innovative approaches to community development (Jacobs, 1989b). In Britain there was a need to enhance non-profit resources to achieve this particularly where organisations were not capable of easily assuming a service delivery role.

Wilson (1989) suggests that the local authority funding of charities in the UK was declining in real terms at the end of the 1980s despite the early gains. Table 10.5 shows the situation regarding central government grants. Table 10.4 indicates that there was an estimated cash increase in local authority payments (grants and fees) to the voluntary sector during the 1980s which partly reflected the increasing contracting out of services to be run on behalf of local government by non-profit organisations. Despite this, at the end of the 1980s non-profit organisations were faced with a real (inflation-adjusted) reduction in funds going to them (Wilson, 1989).

Table 10.4 UK local authority payments to voluntary organisations 1983/84–87/88 (100% estimates by Charities Aid Foundation (CAF)[1])

	1983/84	1984/85	1985/86	1986/87	1987/88
Total out-turn £000s	351,000	429,227	430,702	402,116	500,468
As a % of all local authority Spending	1.3	1.6	1.6	1.4	1.6[2]
Change in cash terms %	100	122	123	115	143
Total at constant prices £000s (1987/88)	421,727	497,189	463,186	422,302	500,468
Change in real terms %	100	118	110	100	119

Source: Charities Aid Foundation (1989)
Notes: [1] One hundred per cent estimates based on CAF survey generated by assuming expenditure of non-responding authorities was equivalent to the average of those which replied.
[2] 1987/88 local authority spending based on DOE estimated out-turn.

Corporate Support

The 1989 edition of the CAF's *Charity Trends* reviewed the performance of charities and non-profit organisations in a comparative environment looking at Canada, the USA and UK. There is little accurate comparative data available concerning the relationships between non-profit organis-

Table 10.5 UK central government grants to voluntary organisations 1981/82–87/88[1]

Department	1981/82 £000	1982/83 £000	1983/84 £000	1984/85 £000	1985/86 £000	1986/87 £000	1987/88 £000
Agriculture, Fisheries & Food	93	91	99	103	118	172	188
Defence	2,637	2,361	2,644	3,285	3,544	4,826	5,297
Education & Science[2]	12,381	11,175	14,318	16,583	18,089	16,230	7,207
Employment[3]	23,207	26,380	27,720	29,180	30,316	33,294	40,290
Energy	91	134	91	344	865	1,118	1,067
Environment – direct grants	763	976	1,220	2,178	2,566	3,612	6,160
– Urban Programme[4,5]	27,000	37,500	46,500	54,000	76,300	76,011	66,644
Foreign & Commonwealth Office	584	–	748	813	823	881	1,203
Health & Social Security	13,775	15,462	23,123	30,068	32,046	35,076	36,572
Home Office	15,396	15,559	16,890	17,088	18,300	19,541	21,946
Industry	479	660	–	–	–	–	–
Lord Chancellor's Department	529	583	604	669	692	727	712
Northern Ireland Departments[6]	20,398	9,480	12,091	13,711	16,801	14,975	14,264
Overseas Development Administration[7,8]	6,393	9,030	10,772	24,027	31,233	26,823	42,476
Scottish Office[9] – direct grants	5,515	6,500	6,824	7,413	8,247	8,970	9,790
– Urban Programme	3,189	4,897	6,750	10,200	11,700	16,600	20,400
Trade & Industry[10]	4,965	5,600	6,221	7,370	8,063	8,689	9,069
Transport	451	440	509	564	605	636	671
Welsh Office – direct grants	2,448	3,280	3,821	4,615	5,094	8,468	5,530
– Urban Programme	690	862	1,325	2,200	2,375	2,850	3,430
Totals	140,984	150,970	182,270	224,411	267,777	279,499	292,916

Source: Charities Aid Foundation (1989)

Notes: [1]The figures in this table exclude grants made for religious activities, to educational bodies which form part of the established educational system (e.g. adult education, research institutes, independent schools) and to the arts. The 1987/88 figures are taken from the Written Answer in Hansard 16 May 1989, 120–1.
[2]The sum listed for DES grants is lower than in previous years because (1) responsibility for grants to village halls and community centres was transferred to local government; and (2) a number of other bodies included hitherto were no longer classified as voluntary bodies.

[3] The figures for 1982/83 onwards do not include MSC payments.

[4] Includes contributions from Department of Education & Science, Department of Health & Social Security, and Department of Transport.

[5] England only (see also Scottish Office and Welsh Office).

[6] Excludes grants to work preparation units (similar to YTS/YOP grants), grants to private and voluntary sector organisations under the Action for Community and the Community Volunteering Scheme programmes, a grant to the NI Arts Council and grants to housing associations.

[7] Overseas Development Administration's total consists solely of grants to various British charities.

[8] The 1982/83 figure includes Foreign & Commonwealth Office grants.

[9] The 1982/83 and 1983/84 figures are estimates only.

[10] Grants totalling £1.3 million were made in 1983/84 to English regional organisations for promotional activities in the UK and overseas. Although these grants were included in previous years, they are now omitted on the grounds that the bodies concerned are not appropriately described as voluntary organisations.

ations and governments, attitudes towards philanthropy and the role of the private sector. The CAF survey argues that this is the result of a lack of interest by governments and researchers and because there is no satisfactory way of comparing international data compiled using different criteria.

In the UK the CAF based its survey of corporate donations on data collected in a questionnaire circulated to 500 leading companies. Of the nation's largest companies, 207 donated an estimated £142 million in cash and non-cash assistance to charities and voluntary organisations in 1987–88. Cash contributions amounted to £61 million. The British figures compared unfavourably with those for the USA, especially when it was considered that 'contributions by some American companies were as much as the social services, health and education budgets of some Third World countries' (CAF, 1989, p.10).

INNOVATION AND TRADITION

In both the USA and the UK in the 1980s, there was a 'commercialisation' of non-profit organisations in response to government spending constraints (Jacobs, 1989b). In Britain, there was a reluctance to introduce commercial criteria into charities while the legal situation restricted the adoption of some innovative US-style funding campaigns. American non-profit organisations expanded their activities often to develop commercial activities little different to those of profit-seeking companies. Purkis (1985, pp.18–19), in research for the National Council for Voluntary Organisations (NCVO), shows that British charities were more closely tied to a system in which there were suspicions about their intentions and restrictive expectations about their role.

Purkis refers to the less 'traditional' forms of charitable and non-profit activities in the USA. These allowed American non-profit organisations to explore new approaches to community involvement. Innovative approaches were often so effective that they prompted small-business lobbyists to demand tougher regulation of the commercial activities of tax-exempt non-profit organisations.

In Britain, traditionalist conceptions of charitable support and self-help defined the legal setting within which charities operated. One long-standing concern was with the definition of charitable activity which included the widely quoted areas concerning the relief of poverty, the advancement of education, and the advancement of religion or 'other pursuits' deemed beneficial to the community but not covered by other charitable activities. The British courts have decided that the 'other pursuits' criterion does not provide a coverall definition of charitable activity, and the Charity Commissioners have resisted pressures to broaden the definition. A trust must be seen to be 'charitable' in its entirety. It must be beneficial to the

228

community, have a 'public' character and be subject to legal administration if necessary. The 'public' aspect requires that the trust must ensure that people external to it benefit by its activities (Tolley's Charities Manual, 1986).

The British Trustee Investments Act 1961 placed restrictions on the freedom of trustees in matters pertaining to investment. The powers of trustees to invest in certain 'wider range' investments were limited under the Act. Limitations were imposed upon the freedom of action of trustees to invest in new property and items such as interest-bearing securities. Trustees were allowed to invest in 'narrow areas' (such as defence bonds, National Savings Certificates, National Development Bonds and deposits in Trustee Savings Banks) but 'special advice' had to be obtained before 'wider investments' could be made in such areas as unit trusts, building societies, debentures and other securities. All such investments were subject to control under the terms of the 1961 Act (Jacobs, 1989b, p.85).

The Act specified that investments had to be suited to the needs of the charity and that they should be diversified. This was so that charities did not become involved in speculative investment (such as property). In effect, the 1961 Act stressed the benefits of longer term stable investments which were regarded as preferable to those associated with risk-orientated 'commercial' investments. The 'play safe' approach was present in the attitude of the Inland Revenue which viewed the 'trading activity' of some charities with considerable suspicion. Charities had to be mindful that the Inland Revenue was aware of the delicate dividing line between charitable trading and property-dealing activities. Tax-exempt status bestowed advantages on charities, so there was a need for scrutiny of both the trading and promotional activities of local organisations.

The Thatcher government's enterprise policy often conflicted with traditionalism. Thatcher stressed the need for community groups to work with the corporate private sector, to be innovative and creative and to seek private sources of funds. This implied that charities had both to work within the confines of the legal constraints placed upon them and devise novel ways of operating in the market. Such an approach required the use of complex administrative and funding arrangements and the introduction of 'commercial' objectives along US lines. It also implied the use of more effective financial control systems and the introduction of improved monitoring and evaluation of organisational performance.

INNOVATION IN BOSTON

Recent research by myself in Boston, USA into the operations of the city's non-profit human services agencies indicated a growing commercialisation and the use of professional evaluation in non-profit organisations (Jacobs, 1985b). It was found that complex networks can exist in the non-profit sector where agencies agree to work together and where there are interactions between fund providers and recipients. The research revealed the following characteristics in the Boston example:

1 There was a desire to 'collectivise' non-profit activity through cooperation between organisations. Collective action enhanced organisational resources and provided greater opportunities for promoting the broader interests (and 'empowerment') of communities.
2 A concentration upon the role of monitoring and evaluation in agencies was important as a way of reviewing their commercial and marketing opportunities (Barry, 1985).
3 Organisations were concerned with adapting to a competitive situation in which competition from the for-profit private sector was seen as a prime consideration.

The agencies surveyed in 1984 provided health care and support for groups including the poor, aged and disabled. The research concentrated upon the private non-profit provision of human services as opposed to services provided by federal and state governments and for-profit companies. Services such as home health visiting, mental health and after care, and services to the blind and handicapped were included. In social services, as opposed to health-related services, there were programmes to assist refugees, samaritan services, rehabilitation services for former prisoners, care programmes for neglected children and a variety of other services.

The delivery agencies had the common link of being in receipt of funds from the United Way of Massachusetts Bay (UWMB) which was directly affiliated to the National United Way of America. The research concentrated upon the UWMB's attempts to encourage local agencies to develop approaches to providing, monitoring and evaluating their services. The UWMB acted as both a central agency providing funds to local affiliates and as a managerial innovator.

Different marketing and evaluation strategies co-existed within the broad framework provided by the UWMB. The UWMB's administrative structure facilitated two fundamental aspects of human service provision: the allocation of charitable funds to local agencies and the subsequent delivery of services to the communities served by those agencies. A degree of centralisation thus existed whereby the UWMB tended to be regarded as a provider of funds and purveyor of agency activities. The UWMB's success in promoting the efficient distribution of resources was made difficult by the size of its overall operation and the diversity of human services for which it catered. During 1984 the UWMB allocated over US$26 million to 183 agencies. Funds were donated by individuals, companies, foundations and others to aid the UWMB's voluntary efforts.

This formed part of the national United Way's overall strategy within its Corporate Development Programme. The programme encouraged private sector bodies and corporations to become actively involved in philan-thropic giving. The major objectives of the national programme concerned the strengthening of company efforts to increase employees' individual giving of contributions, the encouragement of executives and employees to do voluntary work and the fostering of increased corporate involvement in local programmes (United Way of America, 1983, p.5).

In Massachusetts the UWMB at that time executed this strategy in partnership with the United Community Planning Corporation (UCPC). The UCPC acted as the UWMB's 'planning partner' and was directly funded by the UWMB. The UCPC encouraged local agencies to improve the quality of their services and carry out management and programme evaluations. The UWMB/UCPC arrangement reflected a well-defined functional differentiation. The UWMB was concerned with the allocation of resources and the assessment, management and prioritisation of local agency spending. The UCPC was involved with the effectiveness of service delivery and managerial responses to local needs (UCPC, 1983, p.4).

A broad definition of 'evaluation' was adopted by the UWMB and its affiliate agencies. 'Programme evaluation' was the systematic examination of specific programmes 'to provide information on the full range of the programme's short and long term effects' (Hatry, Winnie and Fisk, 1981, p.4). Programme evaluation could include consideration of workloads, operating procedures or staffing, together with a major focus on measuring a programme's impact. This kind of evaluation was identified in many of the UWMB agencies studied.

The UWMB's day-to-day evaluation of allocations of money to agencies involved a form of 'process evaluation', focusing upon the internal proce-dures, staffing and management of programmes, with less emphasis upon programme impacts (Hatry, Winnie and Fisk, 1981, p.4). This was an important form of assessment which stressed local agency dependency upon the UWMB.

In practice, the distinction between programme and process evaluations is often blurred (Anderson and Ball, 1978), particularly when agencies apply a range of methodologies under different circumstances and cater for differing agency requirements. While the programme/process distinction provided a convenient way of describing two general evaluation strategies, it did not reflect the richness of the diverse methodologies applied in the real world of community service provision. It underplayed the importance of the co-existence of approaches which combined elements of each of the two types of evaluation. The UWMB's approach to allocating its funds to agencies exemplified this combination despite its generally process-orien-tated approach. Evaluation therefore involved the following:

1 Resource-focused evaluations with a 'process' orientation (as in the UWMB's fund allocations approach).

2 Client-focused evaluations which often contained a process element (the UCPC's approach, being less important for agencies in their relationship with the UWMB than the resourced-focused evaluations).
3 Evaluations designed for particular local agency or client needs combining 'programme' and 'process' elements.

The UWMB procedure for allocation funds had to operate within clearly defined guidelines where criteria could be developed by which agency objectives could be measured. The UWMB promoted the setting of clear objectives by funded agencies. Objectives were intended to guide the practical implementation of local programmes. The UWMB regarded this as a way of satisfying its own strategic objectives and of meeting the main elements of its mission, as set out below

1 To provide a process for generating the resources necessary to support and foster private voluntary agencies responsive to human service needs;
2 to carry out the process by enhancing the cooperation of contributors and agencies; and
3 to bring together the diverse elements of the Massachusetts Bay Community, and involve leaders in identifying needs and resources (UWMB, 1981).

This required the UWMB affiliate agencies to be accountable and to cooperate with the UWMB in achieving its objectives. The UWMB evaluated programmes as part of a process by which fund receiving agencies were obliged to report to the UWMB on an annual basis. This allowed for an assessment of their records in spending the funds provided by the UWMB and other grant providing agencies. The reporting procedure was coordinated by the UWMB Allocations Coordinating Committee which supported citizen review committees responsible for reviewing local agency performance and activities.

The UWMB/UCPC approach was regarded by some smaller agencies as an acceptable model, but with larger agencies practices tended to reflect special needs, or were determined by the influence of expert evaluators operating within agency national offices. Local agencies reported to the UWMB under the allocations procedures, and this provided a more uniform approach as far as financial accountability was concerned. However, it was necessary to examine how agencies tended to diverge from the UCPC programme 'model' and how this related to some of the constraints and problems faced by agencies.

The American Cancer Society (ACS)

The ACS operated in partnership with the UWMB which provided about 20 per cent of its funding. Funding also came from foundations and corporations. The ACS in Massachusetts worked as a division of the national American Cancer Society which influenced the setting of the local aims and objectives of the ACS. This related to its work in the fields of cancer prevention, research, detection and treatment (ACS, 1976, 1983a). The ACS was a highly professional organisation with an awareness of the need to market its services and raise funds in a competitive environment.

Within the ACS local (or divisional) structure, attention focused upon small ACS units of which there were forty-six covering Massachusetts. These were certified by the ACS. The intention was to provide 'an objective and on-going analysis' of local units (ACS, 1983b, p.55) so that findings could be discussed by unit staff with a concentration on unit effectiveness, structural arrangements (organisational matters) and adherence to national ACS and divisional requirements. Each unit maintained its own goals and objectives, but these could be monitored at divisional level and modified if found to be unsatisfactory.

While this facilitated the meeting of the ACS's own national objectives, the UWMB required the ACS to submit an account of its budgetary and financial objectives. For the ACS, this task was linked to the production of programme activity reports which were produced for national ACS assessment. This entailed a programme by programme evaluation of divisional activities. The ACS was also involved with staff evaluation (for example, with respect to success in fund raising), client evaluation involving an evaluation of client satisfaction, and the evaluation of agencies outside the ACS (a more informal 'evaluation' of suppliers etc.)

This assumed a flexible approach to service delivery and marketing, with evaluation being resource and management-focused rather than client-focused. Concern with the ACS management meeting national objectives, and the use of evaluation to satisfy the requirements of the UWMB and national ACS auditors, underlined this characterisation. The UCPC's client-focused evaluation thus appeared to have little impact within the local ACS.

Boston Children's Service Association (BCSA)

Only about one-eighth of the BCSA's income came from the UWMB (US$251,792 in 1983). This gave the BCSA greater latitude than many of the more dependent agencies within the UWMB orbit. The BCSA could, therefore, develop its own priorities relating to programme funding. However, other external agencies which provided funds demanded scrutiny of the BCSA's activities. The Massachussetts Departments of Social Services and Mental Health were major contributors of funds (nearly two-thirds of the BCSA's income between them) to programmes concerned with family counselling, infant care, post adoptive advice and aid to ethnic minority families (BCSA, 1983a).

The BCSA's approach to evaluation of these programmes was described by a senior BCSA official as 'ad hoc', but related to the objectives of the programmes administered by the BCSA. Despite this approach, the BCSA had a strong commitment to clear objectives for its service administrators and held clear views as to what social needs it was important to cater for within local communities (BCSA, 1978, 1983b, 1984).

In 1983 the BCSA evaluated the impact of charitable private financial and other contributions to the agency's programmes. This involved an assessment of the agency's long-term view of its activities and the establishment of a long-range planning committee to monitor outcomes of planning and finding decisions (BCSA, 1983a). This introduced a more concerted and more coherent approach to evaluation. It went beyond the BCSA's more 'traditional' ad hoc approach and away from a straightforward auditing of reports which had hitherto substituted for proper evaluation. As with the ACS, the UWMB/UCPC did not influence the BCSA's evaluation. Indeed, the UCPC's role was not seen to be particularly relevant to the BCSA's needs. One underlying reason for this was BCSA's non-dependent financial relationship with the UWMB, but there was also an implicit divergence between the pragmatic evaluation adopted by the BCSA and the more structured client focused methodology propounded by the UCPC.

The Medical Foundation

In contrast to the BSCA, the Medical Foundation had a close relationship with the UWMB which contributed a large proportion of the agency's income (over 40 per cent of the Foundation's revenues). The UWMB thus had more influence over the Foundation's activities. A senior Foundation official explained that since the agency relied heavily on the UWMB it was particularly concerned with the UWMB's priorities stated in its Allocations Emphasis Project. The Foundation's central concern with the promotion of medical research through the granting of fellowships to scientists, doctors and medical teachers, fell under the UWMB's low prioritisation category D, or 'less than normal emphasis for growth'. The Foundation was responsible to the UWMB's Health and Rehabilitation Services Committee under its citizen review and allocations procedure, and this was seen to be important for the agency which had to justify rigorously its use of funds within the ambit of the UWMB priorities criteria.

In 1984 the Foundation established a committee on evaluation to conduct internal evaluations for assessing the delivery of services. Emphasis was placed on community health. The Foundation's approach to health promotion concentrated upon how programmes were functioning and attaining their goals, what effects the programmes had upon participants and their impact on target populations (The Medical Foundation, 1984).

The Foundation employed an advisory committee and 'task forces' to analyse client needs. These made recommendations relating to prevention programmes for the aged. The committee encouraged the agency to involve many outside organisations, both public and private, in programme evaluations and in the development of new services to help the agency's clients. The move towards this involvement with other agencies, and the concern to attract a greater proportion of its funds from agencies outside the UWMB, marked an important refocusing of the Foundation's funding and evaluation strategy. This was crucial to overcome the constraints imposed by category D status. In this respect, evaluation served the purpose of defining an overall forward planning strategy for the agency as well as satisfying the basic needs of its clients.

Brookline Community Mental Health Center (BCMHC)

The BCMHC was a small agency, dependent upon the UWMB and the local government authority of Brookline for funding. The UWMB affected the direction of the BCMHC in a manner similar to the Medical Foundation. The UWMB encouraged the BCMHC to implement long-range planning through the citizen review process. The BCMHC also conducted evaluations internally and produced reports for the Massachusetts Department of Mental Health which involved the agency in a partnership implementing mental health programmes.

The BCMHC's approach to evaluation had a strong client focus which was reflected in the BCMHC community mental health needs assessment study (BCMHC, 1983). The study assessed the nature of mental health needs in Brookline to assist in improving services. Internal consultation was initiated with BCMHC staff and with local people in Brookline. The study found that people regarded the BCMHC as providing a good service. This, together with the survey of local needs, helped the agency to improve its services by developing more appropriate objectives.

As with the Medical Foundation, the BCMHC considerd ways in which it could raise funds from alternative sources. Competition was increasing from private mental health practitioners, health maintenance organisations (HMOs) attached to large corporations (providing health care to employees) and from other private health care agencies. Services which could generate revenues were therefore becoming more attractive to the agency despite the numerous restrictions against charging for services under the terms of Massachusetts state partnership contracts.

Visiting Nurse Association of Boston (VNAB)

In contrast to the BCMHC, the VNAB provided an example of a larger non-profit organisation with a very highly developed and sophisticated internal management system. The VNAB had a gross income of US$12 million in 1982, 89.9 per cent of which came from government agencies and only 3.6 per cent from the UWMB (VNAB, 1982a).

The VNAB was aware of the need to compete with profit-seeking companies and to develop an efficient management capability to meet their challenge. With this in mind, the VNAB completed a wide-ranging management review in 1980 which led to the appointment of a new management team and comprehensive internal reorganisation. Reorganisation was accompanied by the redefinition of objectives for agency programmes, the development of effective internal planning mechanisms, the monitoring of management performance and the devolvement of decision-making to the 'lowest appropriate level' (VNAB, 1982b). The agency liaised with the UCPC in obtaining information necessary for market studies designed to orientate towards income-generating services.

Managers completed self-evaluations by rating their own performance and assessing individual contributions to the VNAB. Self-evaluation was 'one of the most important management tools available to a supervisor to utilise in guiding a person towards development of their management skills' (VNAB, 1982c). Management skills were assessed to achieve quality assurance in agency programmes where outcomes were measured against the realisation of management objectives. This approach enabled the VNAB to pay its managers on a competitive incentive basis which was in line with agency plans to adopt a new corporate status which would enable it to enter into profit-making service provision. This involved organising a for-profit 'parent' corporation working in association with a VNAB foundation (non-profit) to channel funds into the parent body.

The VNAB's objectives were constantly under review and the evaluation process was regarded as reflecting the entrepreneurial spirit of the management team. This led to recommendations for organisational, managerial and financial changes which strengthened the agency's move towards the commercialisation of its services.

Commercialisation

In agencies such as the VNAB, objectives were centred upon financial criteria which stressed cost-effectiveness as a way of competing with profit-making organisations. This led to a stress upon the generation of economic surpluses in programmes which previously might have operated on a break-even basis. It was debatable as to whether the UCPC's client focus was the most appropriate in cases where income generation and the earning of surpluses was becoming so crucial. In this respect, the UWMB and other human service agencies were faced with a fundamental choice about their future criteria for measuring agency effectiveness. That choice, particularly for the larger agencies, was essentially about whether human services should run more along commercial lines or whether they should remain in the 'voluntary' sector.

Empowerment Through a Community Foundation

Outside the UWMB, there were other types of non-profit agencies. These illustrate the diversity of the non-profit sector and the potential of agencies in promoting innovative programmes.

At the time of the research, the Boston Foundation was greater Boston's community foundation. It was one of about 300 community foundations located throughout the USA. The Boston Foundation aimed to strengthen the local community in association with other organisations in a common effort 'to heighten the national sense of community' (Boston Foundation, 1986, p.5). The Foundation's grant-making activities depended upon the generosity of donors which comprised both individuals and institutions contributing funds either for general purposes or specific projects. Programmes covered education, health, community development, environmental preservation and the arts. In 1985-86, the Foundation had assets of approximately US$175 million and made grants of nearly US$9 million.

A Trustees' Committee, comprised of representatives from five trustee banks (Bank of Boston, Bank of New England, Boston Safe Deposit and Trust Company, Shawmut Bank of Boston and the State Street Bank and Trust Company) supervised the management of funds while an eleven member Distribution Committee set the policy on grants and their distribution. The majority of the committee was appointed by senior public officials including the Chief Justice of the Supreme Judicial Court of Massachusetts, the Attorney General of Massachusetts, the senior Judge of the US District Court for Massachusetts, the presiding Judge of the Probate Court of Suffolk County, the Chief Justice of the Municipal Court of Boston and the chief volunteer officer of the United Way of Massachusetts Bay. The remainder were appointed by the Trustees' Committee.

The Foundation's sense of community and concern for social justice featured in its 1985–86 *Annual Report*. The strong social orientation of the foundation was of particular interest in the light of its direct links with what some critics called 'the Boston establishment' (Boston Urban Study Group, 1984). The Foundation desired to bring together the 'caring and constructive elements' in the community to serve as a catalyst for individuals and groups to achieve the Foundation's aims. The approach was 'open and responsive to new ideas and arrangements' and conducive to a participatory style. The Foundation used terminology that was reminiscent of some of the more radical community organisations that had been active in Boston at the end of the 1960s:

The effort to counter powerlessness – to resist 'being shifted from one line to another', and to bring form to life that 'seems to have no real order' – is visibly being joined, and defined, by a large and ever-growing number of persons in the Greater Boston area and across the Commonwealth of Massachusetts. Centred in pockets of poverty, where the birthright of equality remains a long way out of reach, it is taking many different forms and has many different expressions. At its core, however, is the attempt to give people a greater voice in decisions that affect their lives and to provide supports and education that will increase their chances of employment. This movement – to release the 'giant' inside and to seek 'a point where fear is no longer a word in my vocabulary' – represents an organised and purposeful quest for empowerment.

(Boston Foundation, 1986, p.9)

There was a quotation in the *Annual Report* from Sylvia Mason's *Flowers in a Field of Thorns* where she called for 'power – inner power and strength' to advance what was called the 'people's struggle in the face of adversity to gain control over their own lives and to shape the kind of future they believe in' (Boston Foundation, 1986, p.9).

The realisation by the corporate private sector that urgent action in the cities was necessary as a way of overcoming the problems of the poor was thus reflected. It provides an example of old-style charity giving way to the spirit of 'community', the search for effective partnership and the legitimation of activism on behalf of the 'powerless' and the social 'underclass' (Murray, 1984).

INNOVATIVE APPROACHES: BRITAIN AND THE US INFLUENCE

In Britain, the marketisation of services and the competition in provision involved the following:

1 American-style approaches to quality assurance, monitoring, evaluation and the marketing of services were of growing importance to British non-profit organisations during the 1980s and 1990s.
2 As in the USA, British non-profit organisations could more effectively operate by linking with other organisations of a similar type. 'Umbrella' organisations such as Business in the Community (although widely differing from the American United Way) promised to bring greater collective strength to local organisations.
3 In a variety of organisations in the UK, there was increased interest in management issues especially as they became more important during a period when central government was placing financial constraints upon local organisations.

4 As in the USA, there was an expansion in the role of non-profit provisions in health, social services, housing and education where government cuts had often adversely affected local service provisions.
5 During the 1980s, there was interest expressed in the UK in US non-profit organisations which extended to concern with the activist nature of local organisations such as the Boston Foundation (Jacobs, 1989).

Collectivising Effort: Business in the Community

Chapter 9 relates how Business in the Community (BITC) promoted corporate social responsibility and urban regeneration through private initiative (Clark and Davies, 1990). BITC was a supporter of a wide variety of independent non-profit organisations involved in urban renewal and community development (BITC, 1989). BITC itself was constituted as a 'not for profit company' limited by guarantee. The UK Charity Commission advised BITC that it would extend recognition to the organisation on the basis of its work to advance industry and commerce by promoting employment opportunities. In this respect, BITC's activities were seen by the commission to be of 'public benefit'.

BITC, in its support role to its affiliated organisations and other groups provides a unique service to non-profit organisations. It attempts to coordinate their activities and lend assistance in the creation and operationalisation of partnerships involving non-profit initiatives. In this way, BITC promotes cooperation within the non-profit sector and seeks to establish links between cooperating local groups and the public and private sectors. There is thus *some* similarity between the role of BITC in relation to its affiliates and associated organisations and that of the United Way in the USA.

BITC has a wide functional role in that it supports organisations in very diverse fields and with broad social, economic and even political aims. BITC's approach is one that translates the concept of business leadership (see Chapter 9) into one that is relevant to the non-profit sector. This requires the active support of the corporate private sector in the development of corporate social responsibility which will assist with the development of community self-activity.

The Per Cent Club

Encouraged by BITC, the Per Cent Club has gathered 300 members, including 72 of Britain's top 200 companies, to promote the aim that companies should contribute not less than half of one per cent of UK pre-tax profits to the community (or one per cent of gross dividends if the company prefers). The club exists to market the idea of the involvement of companies in community and charitable affairs and its existence encourages others to adopt similar principles. In this way 'it has a multiplier effect on the contributions of the private sector' (Weinberg, 1990, p.8).

The aim is to establish regional per cent clubs – two having been launched in 1989 in Sheffield and the north east of England. These attracted over fifty members each immediately after their formation. Regional clubs are regarded as self-standing organisations which adapt their activities to local requirements. Their members include the local offices of national companies that are themselves members of the national Per Cent Club and also local companies which may not have a national profile.

During 1990, the Per Cent Club developed a set of guidelines on corporate community involvement prepared in conjunction with Pannell Ker Foster and the Charities Aid Foundation. Weinberg saw these guidelines as important because they:

> facilitate the definition, calculation, costing and reporting of all community giving and involvement. In particular, the guidelines will address the grey area between community support and commercial operations. This should help introduce greater consistency into existing reporting, as well as encourage many more companies to calculate and publicise their contributions of both cash and non-cash resources.
>
> (Weinberg, 1990, p.10)

Monitoring and Evaluation

BITC identified a need to monitor and evaluate the outcomes of private sector interventions in non-profit initiatives. This was seen as promoting 'good practice' at the local level and contributing to a fund of knowledge about how local initiatives could be made successful. This concern with evaluation was reminiscent of the approaches used by American non-profit organisations and corporate sponsors in their monitoring procedures. It marked an increased concern about the 'productiveness' of corporate investments of cash and non-cash resources in local initiatives and a desire to ensure that there would be measurable 'spin-offs' for the companies involved. This was clear in the case of the United Way and its affiliate human service agencies and also with BITC. All were concerned that private sector investments should be attractive to corporate sponsors and that local programmes were cost effective and of longer term social and economic benefit to companies and the community.

Housing, Urban Renewal and 'Empowerment'

The emphasis upon productiveness, evaluation and monitoring indicates that local initiatives can be regarded in a broadly 'commercial sense'. The borrowing of the American concern about 'outcomes' is therefore in evidence in the BITC strategy (Lord, 1989, p.3). Peter Kenyon (1989), leader of the Finsbury Park Community Trust in north London, describes how a local community can adopt a free market approach in a local partnership involving the private sector and non-profit organisations. The Finsbury Park approach combines the entrepreneurialism of the American strategy with an activist role which attempts to identify community interests with the interests of the corporate private sector (creating a 'shared vision' – see Chapter 9).

The Finsbury Park Community Trust is a pioneering local charity dedicated to the regeneration of the inner city and the promotion of job creation and training for local people. The Finsbury Park initiative was a 'flagship' project for BITC which provided BITC with a fund of information about how similar initiatives could be operated elsewhere.

Kenyon is enthusiastic about the way the local community joined with the private sector to develop a community development plan that was responsive to the needs of the locality and supported by the majority of Finsbury Park residents. This implied an active advocacy role for local non-profit organisations along the lines of the US community initiatives. In such situations, the private sector is not seen as having divergent or antagonistic relations with the local community, but as working in its interests. This may involve the encouragement of local authorities and companies to develop 'pro-community' development proposals or the abandonment of unpopular plans.

Kenyon describes the activist role:

> The threat of an unsuitable development still hung over the area [a site owned by British Rail]. Could the Trust prepare a community based development proposal? Local people clearly wanted affordable housing, some form of development which maximised jobs, and good amenities on the ten acre site. Could a small local charity mobilise the resources for such an outrageously ambitious venture, in an area where land sells for over £1 million an acre?

The answer was 'yes', they could:

> Staff from the Trust, the local Finsbury Park Action Group, and BITC, began to pull together help. Planning Aid for Londoners (a group providing free technical advice) and Hunt Thompson, a firm of community architects, were enlisted. Local housing associations were drawn into an informal consortium.
>
> (Kenyon, 1989, p.2)

A structure was proposed which would enable all these interested parties to work together. In April 1988, a 'Planning for Real' exercise was conducted in order to ascertain local views about the potential uses of the development site. Outline plans were produced and backing for a Finsbury Park Community Plan was sought from the London Borough of Islington.

The plan proposed that 220 houses be constructed on the site, of which 70 per cent would be rented or sold on a shared equity basis. There was also to be 100,000 square feet of workshop space for small businesses and a one-acre garden centre with places allocated for trainees to learn about horticulture. The plan was supported by local councillors, but there were doubts as to whether sufficient funding could be found to implement it. The Trust, however, had interested the Unity Trust Bank (the investment arm of the Co-operative Bank) in the scheme, together with the Housing Corporation and other financial institutions. The formation of a joint venture company was also proposed in which both the investors and local community would have directors.

For BITC, the Finsbury Park initiative was important because it raised a wider issue about the role of private initiative in the regeneration of public housing estates. The Thatcher government's desire to sell off council-owned housing and give control of large estates to former tenants of local authorities was a policy that was pursued with vigour. It was, therefore, an area in which there was likely to be official government encouragement for non-profit housing initiatives, particularly if they were backed by organisations such as BITC.

Ferguslie Park in Paisley, Scotland, is an estate of 6,000 people of which over 1,000 were unemployed in 1989 (BITC, 1989, p.1). The Ferguslie Park Partnership was formed in 1988. The partners included the Scottish Office, the Scottish Development Agency, Strathclyde Regional Council, the Training Agency, Scottish Business in the Community, Renfrew District Council, local residents groups, the local enterprise trust and others. The Partnership developed a strategy which aimed to regenerate the estate over a ten-year period and to combat poor housing, poverty, unemployment, environmental blight and the generally poor image of the estate.

> Innovatory methods to implement the strategy have been developed. The local business community have formed a Business Support Group for the estate. This group will play a crucial role in identifying job and training opportunities for local people.
>
> An Enterprise, Training and Employment Company will be set up on the estate. This will operate a job placement service, arrange customised training in the light of local employers' needs, and make training and advice for those wishing to work for themselves more accessible and relevant. As money is spent on housing and other improvements the Employment Company will make plans to ensure that local unemployed people are ready to take up the jobs available.
>
> (BITC, 1989, p.1)

Such approaches were influenced by experience in the USA where the federal government had encouraged a similar privatisation policy for public housing. For BITC, the USA:

> offers excellent examples of business people and community groups working together on major urban regeneration projects. Long term financial support and technical help from the Ford Foundation and the private sector led Local Initiatives Support Corporation has been crucial to the success of many of these projects.
>
> (Lord, 1989, p.3)

Indeed, there was direct contact between British and American business leaders to discuss ways in which such experiences and knowledge about partnerships could be exchanged. Pete Garcia, described as one of North America's 'most imaginative and successful community leaders' (Lord, 1989, p.3) addressed a UK conference organised by BITC to provide information about US initiatives in the housing field. Garcia was the leader of Chicanos por la Causa, a Mexican-American community group famous from the late 1960s for its militant protests on behalf of Mexican-Americans across the south west states of the USA. Garcia's visit thus provided another example of the way in which militant empowerment advocates were now working closely with the corporate private sector.

Employment Initiatives

Business in the Community also looked abroad to examples of good practice in the promotion of job opportunities. This was a particularly important area of concern during times of economic recession, and it was a field in which non-profit organisations could play a central role. Again, the US influence could be seen in the promotion of partnerships. However, the concern was to take note of developments over a wider international field and to assess the implications of community development, training and job creation initiatives in the USA and elsewhere.

In the domestic context, there were a number of strands of BITC's work which came together in the topic of job creation and employment. BITC was concerned to stimulate local initiative to promote community development not only by encouraging the development of small enterprises, but also by way of improved training, marketing and the assurance of equal opportunities. Large companies were encouraged by BITC through the Professional Firms Group to donate their expertise, skills and advice to community projects (BITC, 1990b, p.10). This involved the provision of expert advice covering diverse areas such as surveying, accounting and legal and consultancy services. Advice could also be offered through the local enterprise agencies where small enterprises could obtain information on training, aspects of good practice, service development and marketing.

With a strengthened regional network of eleven local offices and a coordinated portfolio of partnerships and programmes, BITC was able to extend its activities to a larger number of local organisations. The Prince of Wales, with BITC President's Committee, established guidelines for company boards on how to increase the prosperity of communities through mainstream business activities. These guidelines were known as the Agenda for Action and covered investment, marketing and communications, purchasing and subcontracting, employee community involvement, recruitment, education and training, and management development.

At the local level, BITC had a record of involvement with non-profit organisations such as the Action Resource Centre, Instant Muscle, the Prince's Youth Business Trust and Project Fullemploy. All these were concerned with the problems of the unemployed with particular attention being paid to ethnic minority youth.

Project Fullemploy provides an example of how such organisations can gain respect from the corporate private sector and be regarded as legitimate providers of important services and support to communities at the local level. Project Fullemploy was started by two directors from the international money brokers Charles Fulton and Company and was supported by a number of 'blue-chip' City of London companies. The aim was to harness the goodwill and resources of the private sector, in partnership with central and local government, to address the needs of those at the greatest disadvantage in finding employment.

In practice, Project Fullemploy was mainly concerned with the problems faced by blacks and Asians. This was initially influenced by a concern about the small number of blacks employed in City of London companies and financial institutions. However, it was clear that the problems faced by ethnic minorities in employment were not confined to the square mile of the City. Blacks and Asians were at a general disadvantage in the employment market, made worse by their lack of skills and the poor perceptions of minorities held by employers. In 1976 Project Fullemploy launched a series of courses in skills training mainly for black youths from the inner-London boroughs with teaching staff supplied by Barclays Bank. Project Fullemploy expanded its training programmes with the aid of central and local government funding. Projects were initiated in London and provincial centres designed to meet the needs of both the short- and long-term unemployed.

By 1986 Project Fullemploy claimed to be the main agency outside the education system providing this kind of locally orientated skills training. As a registered charity, Project Fullemploy was operating with a staff of 200 people and a budget of £3.5 million of which around a quarter came from corporate and charitable donations (Jacobs, 1989b, pp.104–5). It was such support which helped tide Project Fullemploy over difficult financial times in 1990 and 1991 when there were doubts about the ability of the organisation to maintain itself. Indeed, it was financial adversity of this kind which served to illustrate the dependency of many non-profit organisations on their financial supporters.

CONCLUSIONS

The incorporation of local organisations into partnerships and other arrangements to facilitate urban renewal and community development is one consequence of the closer relationship between the private sector, government and non-profit organisations. The tendency for partnerships to integrate local groups within projects and programmes is evident across the whole spectrum of interactions between public and private organisations.

The relationships between non-profit organisations, government and the private sector involve a diversity of local and national financial and administrative arrangements. It is this diversity which provides initiatives involving non-profit organisations such flexibility and potential for innovation. In addition, diversity implies that local initiatives can be designed to meet the needs of communities and thereby be attractive to local activists wishing to improve community conditions and increase the resources available to a larger number of people.

The networks between agencies and groups are thus complex. The 'policy community' encompassing non-profit organisations in urban renewal and community development is thus irregular and fluid in terms of its make-up. Indeed, there are a multiplicity of 'policy communities' operating in different towns and cities, each having connections nationally and operating their own regulatory approaches as far as monitoring, evaluation and finance are concerned.

NOTE

Representatives of the following organisations were interviewed in connection with the research in this chapter. The interviews were conducted between 1984 and 1991.

The United States

Action for Boston Community Development (Boston, Mass.); American Cancer Society (Boston, Mass.); Associated Grantmakers of Massachusetts Inc.; Boston Children's Service Association; Boston City Council; Brookline Association for Mental Health; Executive Office of Communities and Development (Mass. State); The Hartford Partnership, Hartford Conn.;

245

Independent Sector, Washington DC; Jamaica Plain Area Planning Council (Boston, Mass.); Massachusetts Department of Social Services; The Medical Foundation (Boston, Mass.); Social Policy Research Group (Boston, Mass.); State of Massachusetts Office of Community Action; Trinity College, Hartford, Conn.; United Community Planning Corporation (Boston, Mass.); United Way of Massachusetts Bay; The Urban Institute, Washington DC; Visiting Nurse Association of Boston.

Britain

Charities Aid Foundation; Business in the Community, London; Finsbury Park Community Trust, London; The Forbes Trust; Local Initiative Support (UK), London; National Council for Voluntary Organisations; Policy Studies Institute, London; Peat Marwick Mitchell and Co., Stoke-on-Trent and London.

11

WHITHER THE CITIES?

In both the USA and Britain, the cities are undergoing changes which have intensified the divisions between regions, social classes and racial groups. The cities urgently need to develop new social, economic and political structures to meet the challenges of today and the future. This task is not insurmountable, nor does it require resources that are beyond the reach of national governments and local communities. The elements of appropriate new structures presently exist in innovative programmes, but sustained effort to develop them is required if there is to be an effective response to the urban problems.

Socio-Technological Issues

It has been shown in previous chapters that urbanisation, economic change and the development of complex technologies are producing new spatial arrangements and communications networks. Managers speak to one another across space. Microtechnology has revolutionised production processes, administration and distribution. Architects are assisted by computer-aided design software while developments in aerospace produce high-tech 'spin-offs' which can be used in the home.

While such developments bring benefits, the urban poor may only glimpse the most exciting technological innovations. 'Smart' laser-guided weapon technology is demonstrated nightly on the television as war unfolds, but in the cities the urban fabric suffers from neglect and social problems seem to be intractable.

Urban Lifestyles

This book has been primarily concerned with two industrialised nations, both of which have comparatively high standards of living and which share certain cultural characteristics. In both, the 1980s was the era of individualism. Conservative governments extolled the advantages of market forces as the expectations of consumers were influenced by the lure of progress through personal advancement and the achievement of higher material standards of living.

Media images pointed to a new urban lifestyle. The upwardly mobile social orientation may have been rejected or despised by some, but it provided a locus around which urban lifestyles could be measured. The elite styles of London, Paris and New York provided the benchmarks against which the aspirant standards of the enterprise culture could be measured. Post-modernism in art, architecture and domestic design symbolised the tone of the modern era; of urban man and woman and the professional 'go-getter'.

It has been seen that the market delivered consumer goods and services, but it could also deprive. The enterprise spirit highlighted both the strengths and weaknesses of the 1980s growth sectors. Job losses were experienced in financial services, the property boom came to an abrupt end, leisure and retailing lost its edge and major high-tech companies announced redundancies. The market thus displayed its limitations despite the success in stimulating economic activity and producing a wide variety of consumer goods. The market could create pollution, social inequality and urban distress as well as the opportunities afforded by consumer choice (Porritt, 1990).

EMPOWERMENT

Partly in response to such problems and the lack of market 'trickle-down', new approaches to community renewal were developed which involved local groups in activities intended to improve local services and promote economic growth. These were often concerned with the provision of support services (such as health care, welfare and so on). Many initiatives provided opportunities for communities to redirect power away from government bureaucracies and centres of corporate power so as to achieve a more balanced mix and favour particular minority or social interests. In many instances, there was scope for developing new activities to enrich the political and social bases of communities by giving priority to specifically local objectives. In Europe and North America this was exemplified by the development of networks and organisations involving a diversity of community groups, environmental or green organisations, urban design coalitions and forward-looking public and private agencies.

The organisational and professional development of community organisations was accompanied by the involvement of local people in groups which often built upon well-tried approaches or developed new strategies. In the USA, non-profit agencies and community development corporations played a part in stimulating new ideas and ways of improving communities. In the UK, organisations such as Business in the Community took a lead in assisting community organisations in their efforts to attract resources and develop managerial skills. These all helped to establish a fund of experience which could be used to develop 'good practice' at the local level.

Tenant Power

While such approaches inevitably implied a degree of 'incorporation' (see Chapters 7 and 8), some initiatives attempted to foster new attitudes in communities to enable them independently to assume responsibilities for their own futures without having to depend upon help from city bureaucrats or expensive private consultants. The development of community self-management and control was particularly evident where tenant organisations took charge of issues affecting the running of large public housing projects. An often cited example was that of the Kenilworth-Parkside complex in Washington DC where the activist Kimi Gray helped residents to organise to combat problems linked to poverty, crime and relations with the public housing bureaucracy.

Even so, there were potential problems for local people, particularly where community autonomy was linked to wider national political issues. In the USA, housing activists had supported the idea of tenant management for many years. However, it was the Reagan and Bush administrations which encouraged the sale of public housing to tenants (as did the Thatcher government) in spite of the outcry from liberals opposed to running down the public housing stock.

Kenilworth-Parkside was one of the initiatives strongly defended by conservatives. Jack Kemp long championed the cause of privatisation and local self-management. When Kemp was appointed by Bush to head the Department of Housing and Urban Development (HUD), Kimi Gray and her tenants were branded by some liberals as 'cogs in Jack Kemp's propaganda machine' (Osborne, 1989, p.4).

Conservatives, by supporting the sale of public housing were able to gain the support of many of those who favoured tenant 'empowerment' at the local level. Liberals thus witnessed the tenant control issue being taken over by the right in pursuit of the goal of privatisation. The anti-bureaucrats were now the conservatives. Kemp's policy of creating tenant management corporations, which would manage projects and be involved in activities to create jobs and raise incomes, had an appeal that was reminiscent of the policies advocated by many of the more radical 'grass roots' activists in the 1970s.

All this implied a reordering of the agenda for community empowerment. It involved the re-casting of ideas that once were regarded as the preserve of the left. 'Community control' could retain its populist appeal while now being used to support private ownership. The market approach was favoured by conservatives because it stressed the need for

249

bureaucracies to be limited and for citizens to be regarded as 'consumers' who had rights and who could express legitimate ideas about how local services should be provided.

Attitudes and 'Hot spots'

In Britain, Business in the Community (BITC) was aware of the need for communities to develop positive attitudes about themselves and in the USA the promotion of community-level focus groups had a similar objective. However, the focus group approach has been open to the criticism that it could simply be a way for corporate and foundation sponsors of projects to monitor community attitudes. The evidence suggests that this need not necessarily be the case.

Focus groups usually consist of around a dozen people carefully selected to represent a cross-section of people. They meet with a moderator who asks their opinions on topics of special interest to sponsoring organisations. Businesses have used this approach to measure consumer acceptance of products, but the technique has also been used in community development projects. An example comes from Columbus in the USA where the city enlisted the assistance of Shelley Berman Communications, a Columbus-based advertising agency with a national reputation for focus group research. Three groups were used, divided according to household income and reflecting the demographic composition of the city.

Participants were reported to have been very supportive of the initiative since it provided them with an opportunity to tell the moderators about the state of local services and the fact that they were poorly consulted on service quality issues (Rinehart, 1986). The consensus of opinion was that the city should provide more information about what it was doing and more information about services and council departmental functions. This was despite a relatively good overall record by the city in attempting to communicate more effectively with the public.

It found that citizens in Columbus placed far more emphasis upon the need for good basic services than on the development of growth-related activities. Even though the city was regarded as a fast-growing one economically within the north east, there was a strong desire to maintain high standards in areas such as street improvement so that standards would not deteriorate as expansion continued. So-called 'hot spot' items covered the need to improve the environment and deal more effectively with crime.

This example provides but one way in which consultation with local consumers of services can be improved. While it marks a way forward, its limitations are obvious in that focus groups provide only a limited number of community representatives the chance to express themselves. If a broader range of attitudes are to be taken into account, and citizens given the opportunity to articulate their ideas and develop them, then a more broadly based approach needs to be adopted. Other experiments have been successful in generating community consultation and involvement, especially those that have been tried in the field of planning and community architecture where local people have been collectively involved, together with professional advisors and architects, in the planning of communities and individual dwellings. The important combination of local autonomy and expert professional involvement to maximise community resources is thus underscored. These two elements can help to foster local self-management exercises which are not simply outgrowths of city hall or corporate initiatives but which can both enjoy autonomy and operate in conjunction with public and private agencies on a co-equal basis.

Rudats

The community architecture and community urban design movements signified a desire by communities to articulate their concerns about the future development of neighbourhoods. David Gosling (1988a) points to the broader implications of the rise of the 'community regeneration industry' in Britain and the United States. He refers to the 'Remaking Cities' conference held in Pittsburgh in 1988 which expressed the vision of David Lewis, the founder of Urban Design Associates of Pittsburgh. Lewis was one of the pioneers of citizen participation and advocacy planning in the United States long before the term 'community architecture' became common in Britain (Gosling, 1988a).

Lewis was influential in the establishment of Regional and Urban Design Assistance Teams (Rudats) set up as interdisciplinary task forces through the American Institute of Architects (AIA) to help communities develop action plans. Rudats are initiated once an invitation has been received form a community, city council or local industry, and participation is then organised on a voluntary basis.

'Remaking Cities' was preceded by the first Anglo-American Rudat. Six British participants joined twelve Americans in a Rudat study team for the Monongahela or 'Mon' Valley (AIA, 1988). The location was one that had been a major steel producing area in Pittsburgh but which had declined as the industry ran-down. Therefore, the participants in the Rudat had an opportunity to compare the process of decline in the United States and Britain with reference to South Wales, Sheffield and Cleveland.

Meetings were held between local citizen groups and the Rudat team which was serviced by the AIA. About fifty meetings were held with two hundred local people. A Mon Valley Commission was established which included representatives of the unemployed, the steel industry, business, trade unions, churches, local government, teachers, social workers, training personnel and chambers of commerce (Gosling, 1988c). A preliminary report was published in 1987 as a basis for the Rudat study.

'Remaking the Cities' took the preliminary findings as a starting point. It also provided an international forum for the discussion of issues which were common on both sides of the Atlantic. This followed the 1986 London 'Building Communities' conference which strengthened international links as well as advanced the community architecture movement in Britain. The legitimacy of the Rudat approach (with 98 Rudats operating in the USA by 1988) was emphasised by the participation of the Prince of Wales at the 1988 conference and his public support of 'citizen architecture'.

The 'Remaking Cities' conference was structured around a series of workshops covering new economic opportunities, the nature of the evolving metropolis, the preservation of neighbourhoods, the creation of new partnerships and research into urban futures. As Gosling comments:

> 'Remaking Cities' was superbly organised and produced a refreshing re-examination of urban problems. Set in Pittsburgh, it also produced a vivid comparison between the glittering downtown of corporate headquarters (with buildings by Philip Johnson and Kohn Pendersen Fox) with the collapse of the steel communities in the Mon Valley. It was a mirror image of the [British] north–south divide which, although officially denied, is a reality in Britain.
>
> (Gosling, 1988a, p.17)

Cowan (1988) reports that professional planners in Britain were becoming more enthusiastic about the potential value of Rudats. The supporters of the Pittsburgh initiative were keen to extend the idea into Britain in spite of the different circumstances facing planners there. Manchester was to be the venue of a British Rudat-style exercise to develop visionary proposals for the city and identify sites where environmental art and design could contribute to urban regeneration. However, the Manchester initiative did not replicate the American approach. For it strictly to have done so, the exercise would have had to have been preceded by months of preparatory work involving a range of community interests and professionals. Manchester's preparation was more modest; nor did the initiative come from the local community. It was influenced by professional architects and the city's bid to host the 1996 Olympic Games; hence the title of the event, 'Manchester 1996: An Olympic City'. Even so, the organisers were committed to developing a new vision for the city and the fostering of a new relationship between professionals and the public.

The Birmingham (UK) City Centre Challenge Symposium, known as the Highbury initiative, also set out to create 'new ground-rules for interventionist ideas in the inner city' (Gosling, 1988b, 1988c). The initiative was sponsored by Birmingham City Council and the Birmingham City Action Team and was intended to advise on the future of the city centre. A report was compiled by the architectural firm DEGW (1988) which

252

was one of the co-organisers of the event. The report pointed to the 'city as theatre' in the context that it had to be attractive if it was to be a popular place to live and conduct business. However, for DEGW, the 'city as theatre' invariably acted as a cover for the urban squalor that surrounded the central business district. Cities should not be developed simply to impress passing visitors and international business people.

Participants from as far afield as Los Angeles and Tokyo were invited to contribute their ideas as to how Birmingham could inject a new spirit into its central area and reinvigorate surrounding communities. Ideas included the creation of a media quarter and the development of Chinese and jewellery quarters which would introduce a richer diversity of 'human scale' activities into a city centre blighted by concrete office blocks, the ring road and featureless pedestrian underpasses.

Cowan (1988) is critical of the Birmingham exercise on several counts. He alludes to the fact that the participants were keen to achieve a workable consensus. For Cowan, this could lead to the creation of 'tablets of stone' which would act as the parameters within which planners worked in the future. In addition, there were few women and no ethnic minorities involved in the initiative. Greater participation from these groups may have given the initiative a different view of the problems facing the city and the needs for which planners should cater.

Workplace Democracy

Some people believe that community control is only one element in getting to grips with social and economic issues. The green movement has stressed the importance of tackling pollution caused by industrial process and the need to effect major changes in industry's attitudes on such issues. Jonathon Porritt has called for a re-examination of industry's social objectives and the ways in which workers can be involved in helping to set socially responsible objectives (Porritt, 1990, p.158).

Kirkpatrick Sale (1980) examines the benefits of the introduction of worker participation in different settings. However,

> In countless cases, from small, co-operative health food restaurants to large manufacturing plants, from the Briarpatch Auto Works in Palo Alto to the asbestos mine in Vermont, from the Scott Bader Commonwealth in England to the metalworking factories of Stockholm, workers have found time and again that mere formal ownership doesn't necessarily have much to do with day-to-day authority.
>
> (Sale, 1980, p.366)

For Sale, effective workplace democracy demands that adequate mechanisms be provided so that workers can effect real influence over the company strategies. Forms of common ownership provide only one basis upon which democratic structures can be developed. The diversity of approaches which are available, for Sale, would be both democratically beneficial and conducive to the improvement of the quality of decision-taking on a community-wide level. Even so,

> I think it would be a mistake to assume that worker control and community self-interest will always coincide, as by some communal magic. Enough evidence has come from the Yugoslav experience, as well as from any number of well-intentioned collectives and community development corporations in this country [the USA], to show that there will be times when workers are going to put their own (or their firm's) concerns ahead of the specific considerations of the outside populace.
>
> (Sale, 1980, p.382)

Sale's solution to this problem is for collectively owned and managed enterprises to have a contractual obligation to the communities within which they operate. In this way, they would be legally obliged to return surplus resources or profits to the community. There could also be external community representation on the boards of local companies (as with some community development corporations in the USA) and some means of issuing shares which would be vested in public boards with a community responsibility (reflecting E.F. Schumacher's idea in his 1973 book *Small is Beautiful*).

THE LIMITS TO URBAN GROWTH

The approaches described above illustrate concerted attempts to come to terms with problems of participation, control and empowerment. However, there seems to be little evidence to suggest that low-resource community initiatives can provide much more than partial solutions. Community participation is commendable and essential to a democracy. There are many individuals and organisations with valuable contributions to make to the improvement of urban communities, but this is not enough.

There are larger issues affecting the future of the cities. They are problems which are only partially offset by increased participation at the local level. Addressing global environmental issues and questions concerning the distribution and generation of wealth and the ways in which new institutional mechanisms can be developed are of vital importance, but they require national and international action to be successful.

Governments have tended to define their strategic actions in terms of more investment in infrastructure. In the 1990s, the cities cannot be regenerated by way of 'traditional' approaches without creating intolerable strains on the environment and existing infrastructures. New highway projects contribute to pollution and the destruction of communities.

Underground railways shift congestion to other points on city transportation networks or reinforce the magnetic attraction of existing central business, shopping and entertainments districts. 'Mega' shopping centres dominate city centres and disrupt local communities, but are closed to the public after trading.

Not surprisingly, green and conservationist issues are at the fore in discussions about the urban environment. Cities are regarded by some as places to enjoy. Others see them as places to inhale exhaust fumes. Commercial developments are constructed on the grand scale with little apparent concern for their longer term environmental consequences. The human scale of the cities is neglected in the name of 'renewal'.

Globalisation

This links urban politics to global concerns. For communities, this implies that new forms of organisation will need to be developed which are capable of forcing issues to the fore and effectively dealing with their implications. Localised 'community politics' is not enough on its own to achieve the necessary wider coordination between interests at local, city-wide national and international levels. It is therefore important that interest groups take account of the international repercussions of public policies and the global implications of urban change.

One issue with international implications concerns the pollution generated by the increase in motor vehicles. The motor car has often been seen as the symbol of the modern city and an expression of the right of the citizen of the city to have access to a personal mode of transportation. There is perhaps no better illustration of the high status afforded to the motor car than in what Banham has described as the 'autopia' of Los Angeles where the freeway acts as

> a comment on the sheer vastness of the movement pattern of Los Angeles, but more than that it is an acknowledgement that the freeway system in its totality is now a single comprehensible place, a coherent state of mind, a complete way of life, the fourth ecology of the Angeleno. Though the famous story in *Cry California* magazine about the family who actually lived in a mobile home on the freeways is now known to be a jesting fabrication, the idea was immediately convincing (several other magazines took it seriously and wanted to reprint it) because there was a great psychological truth spoken in the jest. The freeway is where the Angelenos live a large part of their lives.
>
> (Banham, 1987, pp.213–14)

The 1980s saw the continuation of the idea that car ownership was something which was associated with wealth and economic growth. Modern cities were required to adapt to the motor car as one aspect of economic expansion. The super highway reflected the dynamism of the city and the modernity of urban life. The limits to the growth of motor transport were also recognised by those concerned about the future of cities as habitable places. There was a recognition by many public policy-makers that limits would have to be placed on the motor car if 'motorised' cities were not to continue as major contributors to global warming and air pollution.

The problems associated with the continued expansion of city transportation infrastructures were addressed in a report published in 1988 by the Organisation for Economic Cooperation and Development (OECD). The report stated that

Pollution problems are mainly concerned with air pollutants (CO, NOx, HC particulates, lead and so on) and their associated health risks and ecological impacts. Unpleasant odours and soiling are evident and widespread sources of annoyance. The pollution of the marine environment by the routine or accidental discharge of oil products is also a cause for concern.

Noise disturbance, particularly due to road traffic, is an increasingly common nuisance, especially in densely built-up residential areas and at night.

Land consumption by transport infrastructure (airports, harbours, roads and railways, pipelines and associated facilities) may be in conflict with other land uses and also influences access and property values. Obsolete transport facilities (such as canals, railway lines, stations and tunnels and port facilities) give rise to difficult problems concerning the reconversion of derelict land to appropriate alternative uses.

Solid waste problems mainly involve the disposal of earth and rubble during the building of new transport infrastructures, the recovery and recycling of metals from scrapped vehicles and the disposal of certain non-recoverable materials.

Accident risks mainly concern the daily operation of road transport modes. Although disastrous accidents in the air, sea and rail modes occasionally result in heavy tolls, taken together these generally represent only a small fraction of the deaths, injuries and property damage attributable to road transport activities. The risks caused by the transport of hazardous substances (chemical, nuclear) represent an increasing cause of concern because of the potential scale and intensity of the damage.

The consumption of energy resources by the transport sector is also a major concern because of its high dependence on petroleum. The consumption of metals and other non-fuel minerals for the construction of transport infrastructures and mobile equipment raises longer-term issues of resource utilisation and recycling.

Congestion results in loss of time that might otherwise be spent in more pleasurable or productive activities.

(OECD, 1988, pp.26–8)

Collective Action

The magnitude of these problems poses questions about the impact which environmental and other interest groups can hope to make to change public policy priorities. There may be opportunities for community empowerment through innovative approaches to participation, but the global implications of such issues place limitations upon what groups can expect to achieve on their own. There have been some notable success scored by environmental groups both in the USA and in Britain. However, the OECD report indicates that there are a range of major concerns involved in the environmental issues associated with the development of modern transport infrastructures which are likely to get out of hand unless coordinated public interventions are initiated by urban policy-makers.

The problems raised in the OECD report would seem to call for a strategy by community organisations that maximises their *collective* resources at the local level and coordinates their campaigns nationally. Organisations such as Friends of the Earth and Greenpeace have achieved a degree of success in this respect, but to achieve an effective national and international orientation there would still appear to be a strong case for working in conjunction with established political parties.

Despite the problems associated with such an approach, environmental and other groups have developed contacts with the national parties both in Britain and the USA. In Britain, the Labour party has taken on some of the policies advocated by green organisations although in a diluted form. The Conservative party has adopted a green orientation both by pressure of circumstances and by the need to confront the public popularity and strength of the green lobby. The US and UK governments have also initiated numerous policies affecting the environment and have taken international initiatives, most of which have been criticised by green groups as inadequate.

Conflict and Partnership

The globalisation of policy issues and their seriousness places a heavy burden on public and private agencies which seek to address the problems of the cities through partnership. The approaches to participation discussed above rely upon the consent of those involved and the harmonisation of interests. However, globalised issues are not easily interpreted in terms of consensus. There are inherent conflicts in such issues between those who see economic development as a necessary and desirable way of improving the cities and those who regard growth and its consequent problems (pollution, congestion, etc.) as detrimental to the broader social interest. The corporate private sector increasingly favours socially responsible community partnerships, while local interest groups remain unconvinced that the companies are genuinely concerned about the community.

While the environmental problem acts as a unifying element for environmental lobbyists, it is affected by a number of specific issues concerning different interest groups and clusters of public and private sector organisations. Problems concerning the disposal of waste, the planning of new highways, the monitoring of accidents and so on may be linked, but they are also handled in different ways by different clusters of decision-makers. Different groups in society have various vested interests in maintaining the status quo, such as the transportation unions, road haulage lobby groups and construction companies. A key question thus relates to how national and local interest groups can adopt effective strategies to make the maximum impact in the complex public policy process. In some cases, groups have managed to make an impact by gaining access to politicians and officials who have been responsive to the demands placed upon them. In other instances, there has been less success, making it more difficult for groups to make an impression on policy.

The OECD report confronts the conflict of interests by noting the difficulties involved in reconciling social and economic objectives. It may be socially advisable for governments and the private sector to be supportive of policies which restrict car omissions and enhance the quality of the environment, but the costs involved may be in terms of lost development opportunities for companies, a lower rate of economic growth or the incursion of higher costs caused by measures designed to reduce pollution levels. The nature of the issues and the interests involved will therefore tend to define community partnerships. As global issues become more important, there are likely to be an increasing number of conflict-generating issues concerning urban change and development which will not be resolved by way of traditional consensus-building or 'shared visions' of the future (see Chapter 9).

Policy-makers will be obliged to make stark choices between alternatives. The cities can either continue to develop broadly in the way that they have done up to the present, or radical policies will need to be adopted which reverse the adverse affects of growth but challenge prevailing notions of expansion, profitability and growth. The choices will also involve matters concerning the nature of the market and the degree to which the market should or should not be relied upon to influence change and economic growth. The case for government is strengthened by a recognition that global problems of the kind discussed cannot adequately be resolved by leaving things to market forces (see Chapters 1 and 2). Partnerships for global change may thus come to represent the end product of debates about important issues where some interests may not fit neatly into a 'shared' compromise.

The Limits to Growth

The global politics of urban change requires a recognition that there are limits to the growth potential of the cities. The competition between communities for resources will be played-out in a setting which recognises that continued urban expansion cannot be sustained without incurring the most serious consequences. This recognition is not new. It relates to a concern which goes back to the 1960s when the 'limits to growth argument' became popular. The so-called 'Club of Rome' was established at the end of the 1960s as an informal organisation to foster understanding

> of the varied but independent components – economic, political, natural and social – that make up the global system in which we all live; to bring that new understanding to the attention of policy makers and the public worldwide; and in this way to promote new policy initiatives and action.
>
> (Meadows *et al.*, 1972, p.9)

The members of the club had backgrounds which reflected the diverse interests concerned with environmental issues, with international representatives from industry, the academic world and government. A series of meetings produced the 'Project on the Predicament of Mankind' to examine the complex of problems 'troubling men of all nations' such as poverty, degradation of the environment, loss of faith in institutions, uncontrolled urban expansion, unemployment, alienation of youth, rejection of traditional values and the state of the economy. This was seen as part of the 'world problematique', i.e. that environmental problems have the following characteristics:

1 they occur to some degree in all societies;
2 they contain social, economic and political elements; and,
3 most importantly, they are interrelated.

According to the club

> It is the predicament of mankind that man can perceive the problematique, yet, despite his considerable knowledge and skills, he does not understand the origins, significance and inter-relationships of its many components and thus is unable to devise effective responses. This failure occurs in large part because we continue to examine single items in the problematique without understanding that the

whole is more than the sum of its parts, that change in one element means change in others.

(Meadows *et al.*, 1972, p.11)

The world model was developed as a result of research for the club which investigated five major trends of global concern: accelerating industrialisation; rapid population growth; widespread malnutrition; depletion of non-renewable resources; and a deteriorating environment. The problems associated with these were increasing; the amount of increase each year followed an exponential growth path. The consequence was that there were roughly constant *percentage* increases in growth over time, resulting in an acute crisis for mankind as far as the depletion of resources and the ability to feed the world's population was concerned. The research concluded:

We are convinced that realisation of the quantitative restraints of the world environment and of the tragic consequences of an overshoot is essential to the initiation of new forms of thinking that will lead to a fundamental revision of human behaviour and, by implication, of the entire fabric of present-day society. It is only now that, having begun to understand something of the interactions between demographic growth and economic growth, and having reached unprecedented levels in both, man is forced to take account of the limited dimensions of his planet and the ceilings to his presence and activity on it. For the first time, it has become vital to inquire into the cost of unrestricted material growth and to consider alternatives to its continuation.

(Meadows *et al.*, 1972, pp.190–1)

In the 1990s, these words appear far-sighted. National governments have, since 1972, become more aware of the problems and have adopted measures to deal with some of the worst cases of pollution and environmental degradation. Nevertheless, the continued expansion of the cities is still regarded in favourable terms by governments, the private sector and city citizens. Growth is equated with prosperity, progress, status and cultural development.

But the problems faced now are even more pressing. London and New York are competing on an international level with Paris, Frankfurt, Tokyo and the rapidly expanding cities of the newly industrialised countries of Asia. The Third World sees the consumer paradises of the west as its model for growth and prosperity. The post-Communist countries of eastern Europe have taken the West German and American examples as beacons on the road to faster economic development. There is thus the problem of what happens to the world's resources and the environment if 'everyone' wants to emulate the experiences of the major capitalist cities.

There would presumably be more roads to build, more rapid transit lines to construct, more population growth, more McDonalds outlets in Moscow and more consumer goods to fill the shelves of yet more shopping malls.

People are, of course, entitled to demand such things and deserve an improved standard of living. The desire for better consumer goods and an improved overall standard of living is one that is understandable and fully justified. The issue of key concern is how city decision-makers can effectively respond to achieve these things in a manner that will take account of the adverse spin-offs arising from market-led growth.

Policy Issues

Pierre Laconte, in a keynote paper to the Second International Conference on Environmental Future in 1977, raised some of the most pressing policy concerns. Laconte pointed to the dangers of what Doxiadis (1975) had called 'extreme urbanisation' as the starting point for locating the policy-relevant approaches that could be developed in the cities. The problems associated with extreme urbanisation, according to Laconte, could be summarised in the following way:

1 *Population growth* which continues to pose major problems in world cities.
2 *Industrial growth* which has generated the need for space to accommodate its own activity and which has produced a huge complex of administrative activities and the expansion of the service centres of the cities.
3 *Expanding transportation networks* which have undermined the old 'walkabout' or 'uni-speed' cities.
4 *Urban sprawl* which has been facilitated by the development of transport networks.
5 *Increasing mobility* which has also been characterised by improved telecommunications and which helped to disperse settlements.
6 *Rural decline* which was often seen as an inevitable outcome of 'modernisation' and 'progress'.
7 *Estrangement of city dwellers* which has been the consequence of the city becoming a 'melting pot' of 'strangers' where social controls have been replaced by state intervention to check crime.
8 *Functionalist development of cities* where the actors behind city development have often become anonymous and where powerful property developers influence the pattern of growth in the cities.
9 *Technology versus nature* where the technology that has enabled the cities to grow has also made them independent from nature and its cycles.

10 *Dependence of urban populations on technology* to the extent that mechanical failures potentially threaten the operation of large urban systems. This relates to a variety of possible crisis situations in which cities can be affected by power cuts, natural disasters, nuclear accidents and so on.

11 *Urbanisation versus life-support system* which means that there is a grave threat to the life-support system from continued urban expansion.

In each of the above areas there is a need for policies which will provide a coordinated approach to the interrelated problems outlined. Laconte's suggestion was for a controlled approach to the application of 'middle-of-the-road' technology and the encouragement of the development of small communities which would have a closer relationship with nature.

While Laconte's policy agenda was challenging, there were serious difficulties with the small community strategy and the closer identification with nature. There were dangers here that a utopian view of what should be done could produce unrealistic policy recommendations. However, Laconte did provide some useful practical policy options which could be accommodated in a serious attempt to deal with the problems of continued urban growth and environmental degradation. The encouragement of research into the problems posed, the attraction of non-polluting firms to cities, the development of new planning approaches to conserve energy and weld communities, all seem realisable as objectives for urban development into the twenty-first century.

An Alternative?

Any genuinely 'new' politics to address the problems is probably too much for which to ask. The most effective new approaches will be most likely to develop from existing initiatives and experiences. They should be rooted in the need to cater for the basic interests of citizens by devising policies that are practical yet far-sighted enough to achieve innovative changes in the way in which resources are used and distributed. This will require some state involvement where it is necessary to control the activities of the market. It may even require the development and extension of government initiatives into areas which governments are presently unwilling to enter. Governments already take actions that are designed to ensure that the wider community may benefit (such as in the enforcement of planning controls, the redistribution of resources and the implementation of anti-pollution policies).

Politics should *not* be about state action by bureaucratic control from the centre. The thoroughly discredited state planning of the former Stalinist states of eastern Europe should provide ample evidence of the folly of such an approach. Rational policy alternatives can be developed in conjunction with citizens working through institutions which give expression to their interests and demands. Public bodies should be responsive to these demands and should operationalise policy decisions by working in conjunction with community interests in the ways outlined earlier in this chapter. The Rudats, community architecture, new approaches to community participation and partnership, etc. can all provide examples of the diversity of participative structures that can be developed to assist with policy formulation. National and local governments need to be more open to such initiatives, rather than blunting them in corporatist-style civic programmes. There needs to be a recognition that experimentation and diversity is a healthy characteristic and one that is conducive to more than just aimless 'brainstorming'.

Critical Questions

Given these concerns, city planners and policy-makers will have to face critical choices in the future which will demand action if the cities are to remain as habitable places. Will it be possible to develop more non-waste industrial products? Will it be possible to achieve an equitable distribution of scarce resources given the quest for expansion and economic development in a market economy? Can planners ever achieve effective environmental management? Are we prepared to sacrifice certain lifestyle choices in order to save the environment and redistribute resources to those most in need? Whither the life-support system that the city represents? What are the implications of Laconte's view of 'extreme urbanisation' – is this the root of a major future urban crisis?

These questions are urgently in need of answers. If they are not answered, is society destined simply to resort to traditional values as a kind of cultural and socio-political backlash against the crisis of the cities? The rise of post-modernism in art and architecture has signalled the beginning of such a reversion to past certainties in the pursuit of the 'ideal' harmony between man and his social setting. There is a need to ask questions and seek solutions in ways that challenge existing orthodoxies. The city may then be seen as an obsolete way of organising human activity.

REFERENCES

Adams, C.T. (1990), 'Philadelphia: The Private City in the Post-Industrial Era' in Bernard (ed.), pp.209–26.

Aglietta, M. (1979), *A Theory of Capitalist Regulation: the US Experience*, 2nd edn, London: New Left Books.

Alchay R.E. and Mermelstein, D. (ed.) (1976), *The Fiscal Crisis of American Cities*, New York: Vintage Books.

Almond, G. (1988), 'Separate Tables: Schools and Sects in Political Science', *Political Science and Politics*, No.4. Fall.

Almond, G.A. and Verba, S. (1963), *The Civic Culture: Political Attitudes and Democracy in Five Nations*, Princeton: Princeton University Press.

American Cancer Society (ACS): Massachusetts Division (1983a), *Annual Report: Sharing a Healthy Concern*, Boston, Mass.: ACS.

American Cancer Society (ACS): Massachusetts Division (1983b), *Handbook for Unit Presidents and Vice Presidents*, Boston, Mass.: ACS.

American Cancer Society (ACS) (1976), *What it is; What it Does; How it Began; Who Directs it; Where it is Going*, Washington DC: ACS.

American Institute of Architects (AIA) (1988), *Remaking the Monongahela Valley*, Mon Valley Pittsburgh Rudat.

Anderson, S.B. and Ball, S. (1978), *The Profession and Practice of Program Evaluation*, San Francisco: Jossey-Bass.

Anwar, M. (1986), *Race and Politics: Ethnic Minorities and the British Political System*, London: Tavistock.

Ardrey, S.C. and Nelson, W.E. (1990), 'The Maturation of Black Political Power: The Case of Cleveland', *Political Science and Politics*, xxiii(2), pp.148–51.

Ashford, N. and Davies, S. (ed.) (1991), *A Dictionary of Conservative and Libertarian Thought*, London: Routledge.

Ashford, N. (1991), 'Neo-Conservatism', in Ashford, and Davies, S. (eds) (1991), pp.184–85.

Assefa, H. and Wahrhaftig, P. (1989), 'Managing Urban Conflict: The MOVE Crisis in Philadelphia' in Rosenthal *et al.* (eds), pp.255–77.

Atwater Jr. H.B. and Brewster, H. (1982), 'The Corporation As Good Citizen', in Council on Foundations (1982) pp.17–18

Audit Commission (1988), *The Competitive Council*, London: HMSO.

Audit Commission (1989), *Urban Regeneration and Economic Development: The Local Government Dimension*, London: HMSO.

Audit Commission (1991), *Urban Regeneration and Economic Development: The European Dimension*, Information Paper 5, London: HMSO.

Bach, J. (1990), 'Policy Leadership Roles', *Nation's Cities Weekly*, 13 (48), p.12.

REFERENCES

Banfield, E.A. (1971), *The Unheavenly City Revisited*, Boston: Little Brown.

Banham, R. (1987) *Los Angeles: The Architecture of Four Ecologies*, 2nd edn, Harmonsworth: Penguin Books.

Baraka, I. A. (1972), 'Toward the Creation of Political Institutions for All African Peoples', *Black World*, 21 (2), pp.54–66.

Barclays Bank, (1985), *Barclays Economic Review*, August, London: Barclays Bank.

Barclays Bank (1989), *Barclays Economic Review*, November, London: Barclays Bank.

Barclays Bank (1990), *Barclays Economic Review*, November London: Barclays Bank.

Barnes, W.R. and Penne, R.L. (1984), *Employment Problems and America's Cities*, Washington DC: The National League of Cities Office of Policy Analysis and Development.

Barry, B.W. (1985), *Strategic Planning Workbook for Non-profit Organisations*, Developed by the Management Support Services of the Ameherst H. Wilder Foundation.

Bartlett, C. and Ghoshal, S. (1989), *Managing Across Borders: The Transnational Solution*, Cambridge Mass.: Harvard Business School Press.

Bartsch, C. (1985), *Reaching for Recovery: New Economic Initiative in Michigan*, Washington DC: Northeast-Midwest Institute, The Center for Regional Policy.

Batley, R. (1989), 'London Docklands: An Analysis of Power Relations Between UDCs and Local Government', *Public Administration*, 67(2), Summer, pp.167–87.

Beckworth, D. (1986), 'The Outlook for the Bush Years: Reaganism Without Ideology, Persistence Without Brilliance – and Serious Trouble With Congress', *Time International*, 47, 21 November, pp.14–18.

Beetham, D. (1970), *Transport and Turbans*, London: Oxford University Press for the Institute of Race Relations.

Bentham, C.G. (1985), 'Which Areas Have the Worst Urban Problems?', *Urban Studies*, 22, pp.119–31.

Bernard, R.M. (ed.) (1990a), *Snow Belt Cities: Metropolitan Politics in the Northeast and Mid-west Since World War II*, Bloomington and Indianapolis: Indiana University Press.

Bernard, R.M. (1990b), 'Introduction: Snowbelt Politics', in Bernard (ed.) (1990a), pp.1–24.

Bingham, R.D. and Blair, J.P. (eds) (1985), *Urban Economic Development*, Urban Affairs Annual Review, 27, London: Sage.

Birmingham City Council (UK) (BCC) (1986), *1986 Review: Birmingham the Business City, Economic Strategy*, Birmingham: BCC.

Birmingham City Council (UK) (1989), *Annual Report and Accounts*, Birmingham: Birmingham City Council.

Birmingham City Council (UK) (1990), *Annual Report and Accounts, 1989–90 and Priorities '90–'91*, Birmingham: Birmingham City Council.

Birmingham City Council (UK) (1991), *Birmingham City Council Budget, 1991–92*, Birmingham: Birmingham City Council.

Birmingham Inner-City Partnership (ICP) (1983), *Inner-City Partnership Programme 1983–86*, Birmingham: ICP.

Blomeyer, G. (1988), 'Myths of Malls and Men', *The Architects' Journal*, 187 (20)., pp.38–45.

Bolick, C. (1988), *Changing Course: Civil Rights at the Crossroads*, (New Bruswick, Transaction Books.

Borchert, J.G., Bourne, L.S. and Sinclair, R. (ed.) (1986), *Urban Systems in Tran-*

sition, Netherlands Geographical Studies, Utrecht: University of Utrecht Department of Geography.

Bosanquet, N. (1983), *After the New Right*, London: Heinemann.

Boston Children's Service Association (BCSA) (1978), *Agency Self-Study*, Boston, Mass.: BCSA.

Boston Children's Service Association (BCSA) (1983a), *Annual Report*, Boston, Mass.: BCSA.

Boston Children's Service Association (BCSA) (1983b), *Report on Placement Service Goals*, Boston, Mass.: BCSA.

Boston Children's Service Association (BCSA) (1984), *Internal Training Programme: Needs Assessment Results*, Boston, Mass.: BCSA.

Boston Foundation (1986), *The Boston Foundation Annual Report 1985–86*, Boston, Mass.: The Boston Foundation.

Boston Foundation (c. 1987a), *Making a Difference: The Poverty Impact Program*, Boston: The Boston Foundation.

Boston Foundation, (1987b), *The Boston Foundation Report*, Winter, Boston: The Boston Foundation.

Boston Urban Study Group (1984), *Who Rules Boston? A Citizen's Guide to Reclaiming the City*, Boston, Mass.: Institute for Democratic Socialism.

Boutelle, P. (1967), *The Black Uprisings: Newark, Detroit 1967*, New York: Merit Publishers.

Broadwater Farm Inquiry (1986), *Report of the Independent Inquiry into Disturbances of October 1985 at the Broadwater Farm Estate, Tottenham, Chaired by Lord Gifford QC*, London: Kira Press.

Brookline Community Mental Health Center (BCMHC) (1983), *Newsletter*, Summer, Brookline, Mass.: BCMHC.

Brooks, A. (1983), 'Black Business in Lambeth: Obstacles to Expansion', *New Community*, 11, pp.42–54.

Browne, L.E. (1988, originally 1983), 'High Technology and Business Services' in Lampe (ed.) (1988), pp.201–24.

Browning, R P. and Marshall, D.R. (1986), 'Black and Hispanic Power in City Politics: A Forum', *Political Science and Politics*, xix(3), pp 573–75.

Browning, R.P., Marshall, D.R. and Tabb, D.H. (eds) (1990), *Racial Politics in American Cities*, New York: Longman.

Browning, R.P., Marshall, D.R. and Tabb, D.H. (1986), 'Protest is not Enough: A Theory of Political Incorporation', *Political Science and Politics*, xix(3). pp.576–81.

Bruinvels, P. and Rodrigues, D. (1989), *Investing In Enterprise: A Comprehensive Guide to Inner City Regeneration and Urban Renewal (Action for Cities)*, Oxford: Basil Blackwell in association with Brombard Financial Services Group.

Buchanan, J.M., Niskanen, W.A., Roberts, P.C., Stelzer, I.M. and Budd, A. (1989), *Reaganomics and After*, IEA Readings 28, London: The Institute of Economic Affairs.

Buchanan, J.M., Wagner, R. and Burton, J. (1978), *The Consequences of Mr Keynes: An Analysis of the Misuse of Economic Theory for Political Profiteering, with Proposals for Constitutional Disciplines*, Hobart Papers 78, London: The Institute of Economic Affairs.

Business in the Cities (1988a), *Business in the Cities: A Briefing Note*, London: Business in the Cities.

Business in the Cities (1988b), *Business Leadership Teams: A Briefing Note*, London: Business in the Cities.

REFERENCES

Business in the Cities (1990), *Leadership in the Community: A Blueprint for Business Involvement in the 1990s*, London: Business in the Cities.

Business in the Community (BITC) (undated), *Business in the Community: A Guide to Action*, London: BITC.

Business in the Community (BITC) (1986), *Business and the Inner Cities: How the Business Community Can Work With Others to Promote Better Opportunities in Our Inner Cities*, London: BITC.

Business in the Community (BITC) (1987a), *Business in the Community Magazine: Strategies and Approaches for Developing the Local Economy*, Winter, London: BITC.

Business in the Community (BITC) (1987b), *Business in the Community Magazine: Partners in Local Economic Recovery, Job Creation and Training*, Summer, London: BITC.

Business in the Community (BITC) (1988), *Review of the Year 1988*, London: BITC.

Business in the Community (BITC) (1989), *Partners for Living Cities Special Supplement*, London: BITC.

Business in the Community (BITC) (1990a), *Your Business and the Environment: An Executive Guide*, London: Business in the Environment BITC.

Business in the Community (BITC) (1990b), *Business in the Community Corporate Review*, London: BITC.

Business in the Environment (1990), *Your Business and the Environment: An Executive Guide*, London: BITC.

Cabinet Office (1990), *The Enterprise Challenge: Overcoming Barriers to Growth in Small Firms*, London: HMSO.

Calderdale Partnership (1989), *Vision in Calderdale*, Halifax, Yorkshire: Calderdale Partnership.

Campbell, B. (1990), 'Dangerous Liaisons: The Movement Against the Poll Tax', *Marxism Today*, May, pp.26–7.

Cane, A. (1991), 'On the Brink of Disaster', *The Financial Times*, 23 April.

Cardew, R.V., Langdale, J.V. and Rich, D.C. (eds) (1982), *Why Cities Change: Urban Development and Economic Change in Sydney*, London and Sydney: Allen and Unwin.

Carson, R.T. and Oppenheimer, J.A. (1984), 'A Method of Estimating the Personal Ideology of Political Representatives', *American Political Science Review*, 78(1), pp.163–78.

Cason, D. (1980), *Michigan's State Urban Strategies*, Washington DC: US Department of Housing and Urban Development.

Castells, M. (1977), *The Urban Question*, Cambridge Mass. and London: MIT Press, Edward Arnold.

Castells, M. (1978), *City, Class and Power*, London: Macmillan.

Catalano, A. (1991), 'Investment Still on Hold', *Estates Gazette*, 9101, 12 January, p.59.

Cawson, A. and Saunders, P. (1983), 'Corporatism, Competitive Politics and Class Struggle', in King, (ed.).

Cawson, A. (ed) (1985), *Organised Interests and the State: Studies in Meso-Corporatism*, Newbury Park and London: Sage.

Cervero, R. (1989), *America's Suburban Centres: The Land Use-Transportation Link*, Boston: Unwin Hyman.

Champion, A.G. (ed.) (1989a), *Counterurbanisation: The Changing Pace and Nature of Population Deconcentration*, London: Edward Arnold.

267

REFERENCES

Champion, A.G. (1989b), 'United Kingdom: Population Deconcentration as a Cyclic Phenomenon' in Champion, (1989a), pp.83–102.

Charities Aid Foundation (CAF) (1989), *Charity Trends: 12th Edition*, London: CAF.

Cheshire, P.C. and Hay, D.G. (1989), *Urban Problems in Western Europe: An Economic Analysis*, London: Unwin Hyman.

Cheung, S.N.S. (1978), *The Myth of Social Cost: A Critique of Welfare Economics and the Implications for Public Policy*, Hobart Papers 82, London: The Institute of Economic Affairs.

Chisholm, M. and Kivell, P. (1987), *Inner City Waste Land: An Assessment of Government and Market Failure in Land Development*, Hobart Papers 108, London: The Institute of Economic Affairs.

City of Hartford (1987), *The Hartford Partnership*, Hartford Conn.; The City of Hartford and The Greater Hartford Chamber of Commerce.

Clark, C. and Davies R. (1990), *Guidelines on Corporate Responsibility*, London: BITC.

Clark, G. Gertler, M. and Whiteman, J. (1986), *Regional Dynamics: Studies in Adjustment Theory*, London: Unwin Hyman.

Clark, P. and Rughani, M. (1983), 'Asian Entrepreneurs from Leicester in Wholesaling and Manufacturing', *New Community*, 11, pp.23–33.

Clark, T.N. and Ferguson, L.C. (1983), *City Money: Political Processes, Fiscal Strain and Retrenchment*, New York: Columbia University Press.

Clark, T.N. and Inglehart, R. (1989), 'The new Political Culture: Changing Dynamics of Support for Welfare State Policies in Post-Industrial Societies', Paper presented at the Annual Meeting of the American Political Science Association.

Clarke, S. (1988), 'Over-accumulation, Class Struggle and the Regulation Approach', *Capital and Class*, 36, Winter, pp.59–92.

Clark, T.N. (1988), 'Comparing Cities Across Nations: How?, Work in Progress on International Comparisons of the Fiscal Austerity and Urban Innovation Project', Paper prepared for the Conference of the Fiscal Austerity and Urban Innovation (FAUI) Project, Washington DC, Chicago: The University of Chicago, FAUI.

Commission for Racial Equality (CRE) (1985), *Annual Report*, London: CRE.

Commission of the European Communities, Directorate-General Information, Communication and Culture (1991), *Monthly Newsletter on the Single Internal Market*, Luxembourg: Commission of the European Communities.

Commonwealth of Massachusetts, (1983), *The Report of the Governor's Task Force on Private Sector Initiatives*, Medford, Mass.: The Lincoln Filene Center for Citizenship and Public Affairs, Tufts University.

Commonwealth of Massachusetts (1984), *Public Sector Resources for Businesses' Financial Needs: Appendix D, Report of the Governor's Commission on the Future of Mature Industries*, Boston Mass.: State House.

Community Development Project (CDP) (1977), *Gilding the Ghetto: The State and the Poverty Experiments*, London: CDP Inter-Project Editorial Team.

Community Service Volunteers (CSV) (1985), 'Recommendations of A Seminar Held on 16 October Promoted by Watts Labour Community Action Committee to Promote Black Community Enterprise Initiatives', London: CSV.

Confederation of British Industry (1988), *Initiatives Beyond Charity: Report of the CBI Task Force on Business and Urban Regeneration*, London: CBI.

Congressional Quarterly (1978), *Urban America: Policies and Problems*, Washington DC: Congressional Quarterly.

Cooke, P. (ed.) (1989), *Localities: A Comparative Analysis of Urban Change*, London: Unwin Hyman.

Council for Financial Aid to Education Inc. (CFAE) (1986), *Corporate Support of Education 1985*, New York: CFAE.

Council on Foundations (1982), *Corporate Philanthropy: Philosophy, Management, Trends, Future, Background*, Washington DC: Council on Foundations.

Cowan, R. (1987), 'Peace Pact for Royal Docks', *Architects Journal*, 7 October, p.17.

Cowan, R. (1988), 'Rudat Awakening After Pittsburgh', *Architects Journal*, 189, (16), April, p.17

Cowie, H. (1985), *The Phoenix Partnership: Urban Regeneration for the 21st Century*, London: National Council of Building Material Producers.

Crewe, I. (1990), 'Labour's Forward March: Centre of Attraction', *Marxism Today*, May, pp.14–17.

Cummings, S. (ed.) (1988), *Business Elites and Urban Development*, Albany: State University of New York Press.

Dahl, R.A. and Lindblom, C.E. (1976), *Politics, Economics and Welfare*, Chicago: University of Chicago Press.

Dahl, R.A. (1956), *Preface to Democratic Theory*, Chicago: University of Chicago Press.

Dalby, S. (1990), 'High Tech Artery: Growth Industries in the Thames Valley', *Financial Times*, 16 February.

Davies, H. (1984), *1981 Census: A Ward Index of Deprivation*, GLC Statistical Series 35, London: Greater London Council.

Davies, P. (1980), *The Metropolitan Mosaic: Problems of the Contemporary City*, BAAS Pamphlets in American Studies 4, Durham: British Association for American Studies.

Davies, S. (1991), 'The New Right', in Ashford and Davies, S. (ed.), pp.187–9.

Davis, J.A. (1986), 'British and American Attitudes: Similarities and Contrasts' in Jowell, R. *et al.* (1986), pp.88–114.

Davis, M., Hiatt, S., Kennedy, M., Ruddick, S. and Sprinker, M. (eds) (1990), *Fire in the Hearth: The radical Politics of Place in America. Volume Four of the Year Left*, London and New York: Verso.

Davis, P. (ed.) (1986), *Public Private Partnerships: Improving Urban Life*, New York: The Academy of Political Science in conjunction with the New York City Partnership.

Davis, T. (1986), 'The Forms of Collective Racial Violence', *Political Studies*, 34, pp.40–60.

DEGW, (1988), *The Highbury Initiative: Proceedings of the Birmingham City Centre Challenge Symposium*, London: DEGW.

Deane, M. (1990), 'Will the City Survive? *The World in 1991*, London: The Economist Publications, pp.128–30.

Dearlove, J. (1973), *Politics and Policy in Local Government*, Cambridge: Cambridge University Press.

Derbyshire County Council (DCC) (1989a), *Burnaston Diary: Key Events Leading up to Toyota's Announcement*, Information Series, Matlock: DCC.

Derbyshire County Council (DCC) (1989b), *Derbyshire Secures Toyota Plant*, Information Series, Matlock: DCC.

Derbyshire County Council (DCC) (1989c), *Toyota Impact Study: Outline Format*, Matlock: DCC.

Derbyshire, I. (1987), *Politics in the United States: From Carter to Reagan*, Edinburgh: Chambers.

DeWit, J. (1969), *Indian Workers' Associations in Britain*, London: Oxford University Press for the Institute of Race Relations.

Dorfman, N.S. (1988, originally 1983), 'Route 128: The Development of a Regional High Technology Economy', in Lampe (ed.) (1988), pp.240–74.

Downs, A. (1957), *The Economic Theory of Democracy*, New York: Harper.

Doxiadis, C.A. (1975), 'The Ecological Types of Space that We Need', *Environmental Conservation*, 2 (1), pp.3–13

Duckworth, F.C. (1983), 'On the Influence of Debate on Public Opinion', *Political Studies*, 31, pp.463–78.

Duggan, P. and Mayer, V. (1985), *The New American Unemployment: Appropriate Government Responses to Structural Dislocation*, Washington DC: Northeast-Midwest Institute.

Dungate, M. (1984), *A Multi-Racial Society: The Role of National Voluntary Organisations*, London: National Council for Voluntary Organisations.

Dunleavy, P. (1979) 'The Urban Bases of Political Alignment', *British Journal of Political Science*, 9, pp.409–43.

Dunleavy, P. (1980a) *Urban Political Analysis*, London: Macmillan.

Dunleavy, P. (1980b), 'Social and Political Theory and the Issues in Central-Local Relations' in Jones (ed.) (1980), pp.116–36.

Dunleavy, P. (1991), *Democracy, Bureaucracy and Public Choice: Economic Explanations in Political Science*, London: Harvester Wheatsheaf.

Eckstein, G. (1981), 'Supply-side Economics: Panacea or Handout for the Rich?', *Dissent*, 28 (2), Spring.

Edwards, J.E. and Batley, R. (1978), *The Politics of Positive Discrimination*, London: Tavistock Publications.

Eldredge, H.W. (ed.) (1967), *Taming Megalopolis: Volume One, What is and What Could Be*, New York: Doubleday.

Elster, J. (1986), *Rational Choice*, Oxford: Basil Blackwell.

Engels, F. (1962) *The Condition of the Working Class in England*, Moscow: Progress Publishers.

Engels, F. (1979), *The Housing Question*, Moscow: Progress Publishers.

Engstrom, R.L. and McDonald, M.D. (1981), 'The Election of Blacks to City Councils: Clarifying the Impact of Electoral Arrangements on the Seats/Population Relationship', *American Political Science Review*, 75(2), pp.344–54.

Engstrom, R.L. and McDonald, M.D. (1982), 'The Underrepresentation of Blacks on City Councils: Comparing the Structural and Socioeconomic Explanations for South–Non-South Differences', *The Journal of Politics*, 44, pp.1088–99.

Environment, Department of (DoE) (1989), *Progress on Cities*, London: HMSO.

Environment, Department of (DoE) (1990), *USA Experience in Evaluating Urban Regeneration*, London: HMSO.

Environment, Department of (DoE) (1991), *Department of the Environment Annual Report, 1991: The Government's Expenditure Plans 1991–92 to 1993–94*, Cmnd. 1508 London: HMSO.

Estates Gazette (1988), 'Docklands Duo', (editorial), 9 April, p.3.

Estates Gazette (1989), 'Regeneration: Innovators Must be Protected', (editorial), 4 June, p.4.

Fainstein, N.I. and Fainstein, S.S. (1983a), 'Regime Strategies, Communal Resistance and Economic Forces', in Fainstein, *et al.* (eds) (1983) pp.245–82.

Fainstein, S.S. and Fainstein N.I. (1983b), 'Economic Change, National Policy and the System of Cities' in Fainstein *et al.* (eds) (1983), pp.1–26.

Fainstein, S.S., Fainstein, N.I., Hill, R.C., Judd, D.R. and Smith, M.P. (eds) (1983),

REFERENCES

Restructuring the City: The Political Economy of Urban Redevelopment, New York and London: Longman.
Feagin, J. and Smith, M.P. (eds) (1987), *The Capitalist City*. Oxford: Blackwell.
Feder, J. and Hadley, J. (1986), 'Cutbacks, Recession and Hospital Care for the Urban Poor' in Peterson, and Lewis (eds) (1986), pp.37–61.
Federal Reserve Bank of Boston (1986), *New England Economic Indicators*, Third Quarter, Boston: Federal Reserve Bank of Boston.
Financial Times (1988), 'Assessing the Cost of the Cure', 28 November.
Forbes, I. (ed.) (1986), *Market Socialism: Whose Choice?* London: Fabian Society.
Frey, W.H. (1989), 'United States: Counterurbanisation and Metropolis Depopulation, in Champion (ed.) (1989a), pp.34–61.
Friedland, R. (1982), *Power and Crisis in the City: Corporations, Unions and Urban Policy*, London and Basingstoke: Macmillan.
Gamble, A. (1985), *Britain in Decline: Economic Policy, Political Strategy and the British State*, 2nd ed, Basingstoke: Macmillan.
Gilliam, R.E. (1975), *Black Political Development: An Advocacy Analysis*, Port Washington, NY: Dunellen.
Gilmour, B. (1990), 'The Battle for (Higher) Retail Sales', *Estates Gazette*, 9045, 10 November, p.72.
Glazer, N. and Young, K. (1983), *Ethnic Pluralism and Public Policy: Achieving Equality in the United States and Great Britain*, London and Lexington Mass.: Heinemann and D.C. Heath in association with the American Academy of Arts and Sciences, Commission for Racial Equality and the Policy Studies Institute.
Gleeds (1991), *State of the UK Construction Industry*, London: Gleeds.
Gordon, D. (1976), 'Capitalism and the Roots of Urban Crisis' in Alcahy and Mermelstein (eds) (1976).
Gosling, D. (1988a), 'Remaking the Cities the Rudat Way', *Architects Journal*, 187, (11), March, p.17
Gosling, D. (1988b), 'Putting the Show on the Ring Road', *Architects Journal*, 188, (38), September, p.15
Gosling, D. (1988c), 'Revivalist Meetings', *Architects Journal*, 188 (39), September. pp.26–9.
Goss, S. (1988), 'Consumers, Community and Choice', *New Socialist*, 57, October–November.
Gottdiener, M. (1987), *The Decline of Urban Politics: Political Theory and the Crisis of the Local State*, Sage Library of Social Research, vol.162, Newbury Park and London: Sage.
Gould, B. (1989), *A Future of Socialism*, London: Johnathan Cape.
Gray, J. (1990), *Limited Government: A Positive Agenda*, Hobart Papers 113, London: The Institute of Economic Affairs.
Green, D.G. (1987) *The New Right: The Counter-revolution in Political, Economic and Social Thought*, Brighton: Wheatsheaf Books.
Greenwood, J. and Wilson, D. (1989), *Public Administration in Britain Today*, London: Unwin Hyman.
Gudgin, G. and Taylor, P.J. (1979), *Seats, Votes and the Spatial Organisation of Elections*, London: Pion.
Gurr, T.R. and King, D.S. (1987), *The State and the City*, Chicago and London: University of Chicago Press and Macmillan.
Habermas, J. (1973), *Legitimation Crisis*, London: Heinemann.
Hall, A.L. (1989), *Developing Amazonia: Deforestation and Social Conflict in Brazil's Carajas Programme*, Manchester: Manchester University Press.
Hall, P., Breheny, M., McQuaid, R. and Hart, D. (1987), *Western Sunrise: The*

271

Genesis and Growth of Britain's Major High Tech Corridor, London: Unwin Hyman.

Hall, P. and Preston, P. (1988), *The Carrier Wave: New Information Technology and the Geography of Innovation, 1846–2003*, London: Unwin Hyman.

Hall, S. and Jacques, M. (eds) (1989a), *New Times: The Changing Face of Politics in the 1990s*, London: Lawrence and Wishart.

Hall, S. and Jacques, M. (eds) (1989b), 'Introduction' in Hall, and Jacques (eds) (1989a), pp.11–20.

Hambleton, R. (1990), *Urban Government in the 1990s: Lessons From the USA*, Bristol: University of Bristol School for Advanced Urban Studies.

Hamilton, N. and Harding, T.F. (1986), *Modern Mexico: State, Economy and Social Conflict*, Latin American Perspectives Readers, vol. 1, Newbury Park and London: Sage.

Hannay, P. (1988), 'Close to the Edge', *The Architects' Journal*, 187, (20), pp.48–55.

Hansen, J.M. (1985), 'The Political Economy of Group Membership', *The American Political Science Review*, 79, pp.79–96.

Harding, A. (1990), 'Local Autonomy and Urban Economic Development Policies: The Recent UK Experience in Perspective' in King and Pierre (eds) (1990), pp.79–100.

Harrington, M. (1984), 'Introduction', *Who Rules Boston? A Citizen's Guide to Reclaiming the City*, Boston, Mass.: Institute for Democratic Socialism with the Boston Urban Study Group, pp.6–8.

Harvey, D. (1973), *Social Justice and the City*, Baltimore M: John S Hopkins University Press.

Harvey, D. (1988), *Social Justice and the City*, Oxford: Basil Blackwell.

Hatry, H.P., Winnie, R.E. and Fisk, D.M. (eds) (1981), *Practical Program Evaluation for State and Local Governments*, Washington DC: The Urban Institute Press.

Hegedüs, J. and Tosics, I. (1990), 'Moving Away From the Socialist Housing Model: The Changing Role of Filtering in Hungarian Housing' in van Vliet and van Weesep (eds) (1990), pp.242–58.

Heidenheimer, A.J., Heclo, H. and Adams, C.T. (1990), *Comparative Public Policy: The Politics of Social Choice in America, Europe and Japan*, 3rd edn, New York: St. Martin's Press.

Henderson, J. and Castells, M. (eds) (1987), *Global Restructuring and Territorial Development*, Newbury Park and London: Sage.

Henry, N. (1987), *Governing at the Grassroots: State and Local Politics*, Englewood Cliffs, NJ.: Prentice-Hall.

Heseltine, M. (1987), *Where There's a Will*, London: Hutchinson.

Heseltine, M. (1989), *The Challenge of Europe: Can Britain Win?* London, Weidenfield and Nicolson.

Hillson, J. (1977), *The Battle of Boston*, New York: Pathfinder.

Hindess, B. (1971), *The Decline of Working Class Politics*, London: MacGibbon and Kee.

HMSO (1989), *Regional Trends*, 24, edited by Griffin, T, London: Central Statistical Office.

HMSO (1991), *The Citizen's Charter: Raising the Standard*, London: HMSO.

Holliday, I.M. (1990), 'The New Suburban Right in British Local Government: Conservative Views of the Local', Paper presented to the Political Studies Association (UK) Annual Conference, University of Durham.

Holzer, H.J. (1991), 'The Spatial Mismatch Hypothesis: What Has the Evidence Shown?', *Urban Studies*, 28, (1), pp.105–122.

REFERENCES

Hoover, K. and Plant, R. (1989), *Conservative Capitalism in Britain and the United States: A Critical Appraisal*, London: Routledge.

Houlder, V. (1990), 'A Deep Malaise Which May Take Years to Cure', *Financial Times*, 29 October.

House of Commons (HC) (1990), Social Services Committee, *Low Income Statistics*, HC 376.

House of Commons (HC) (1991), Social Security Committee, *Low Income Statistics: Households Below Average Income Tables 1988*, HC 401.

House of Lords (HL) (1991), Select Committee on Science and Technology, *Innovation in Manufacturing Industry*, HL Paper 18-1.

Housing and Urban Development, Department of (HUD) (1979), *Local Economic Development Tools and Techniques: A Guidebook for Local Government*, Washington DC: HUD and US Department of Commerce.

Hoy, J.C. and Bernstein, M.H. (1981), *Business and Academia: Partners in New England's Economic Renewal*, Hanover and London: University Press of New England for New England Board of Higher Education.

Humphry, D. and Ward, M. (1974), *Passports and Politics*, Harmondsworth: Penguin.

Hyink, B.L., Brown, S. and Provost, D.H., (1989), *Politics and Government in California*, 12th ed., New York: Harper and Row.

Independent Sector (IS) (1988), *Research in Progress: A National Compilation of Research Projects on Philanthropy, Voluntary Action and Not-for-Profit Activity*, Washington DC: Independent Sector.

Independent Sector (IS) (1989), *Dimensions of the Independent Sector: A Statistical Profile*, Washington DC: Independent Sector.

Independent Sector (IS) (1990), *Ten Year Report: Independent Sector 1980-1990*, Washington DC: Independent Sector.

Inglehart, R. (1985), 'Aggregate Stability and Individual-Level Flux in Mass Belief Systems: The Level of Analysis Paradox', *American Political Science Review*, 79(1), pp.97-116.

International City Management Association (ICMA) (1989), *Municipal Year Book 1989*, Washington DC: ICMA

Jacobs, B. (1985b), 'Human Services Agencies in Boston', unpublished paper, North Staffordshire Polytechnic.

Jacobs, B.D. (1980), 'Wolverhampton Council for Community Relations: The Pattern of Participation', *Policy and Politics*, 8, pp.401-21.

Jacobs, B.D. (1984a), 'Labour Against the Centre: The Clay Cross Syndrome', *Local Government Studies*, March/April, pp.75-87.

Jacobs, B.D. (1984b), 'Why is Britain Slow to Grasp Ethnic Needs?', *Charity*, 1(12), p.6.

Jacobs, B.D. (1985a), 'Private Initiative and Community Responsibility: A Strategy for the Black Community', *Public Administration*, 63(3), pp.309-25.

Jacobs, B.D. (1986), *Black Politics and Urban Crisis in Britain*, Cambridge: Cambridge University Press.

Jacobs, B.D. (1987), 'Nonprofit Organisations and Urban Policy in Britain and the USA', *Public Policy and Administration*, 2 (2), pp.31-45.

Jacobs, B.D. (1988), *Racism in Britain*, London: Christopher Helm.

Jacobs, B.D. (1989a), 'Ethnic Minority Representation in Britain: Politics and the Electoral System', Paper presented to the 12th Annual Scientific Meeting of the International Society of Political Psychology, Tel Aviv, Israel.

Jacobs, B.D. (1989b), 'Charities and Community Development in Britain' in Ware (ed.) (1989) pp. 82-112.

273

REFERENCES

Jacobs, B.D. (1990), 'Business Leadership in Urban Regeneration: Towards a Shared Vision?', in King, and Pierre (eds) (1990), pp.195–211.

Jenkins, H. (1991), 'Property Development: Restarting the Cycle', *Estates Gazette*, 18 May, pp.76–7.

Jones, G. and Stewart, J. (1985), *The Case for Local Government*, London: Allen and Unwin.

Jones, G. (ed.) (1980), *New Approaches to the Study of Central-Local Government Relationships*, Aldershot: Gower for the Social Science Research Council.

Jowell, R., Witherspoon, S. and Brook, L. (eds) (1986), *British Social Attitudes: the 1986 Report*, Aldershot: Gower and Social and Community Planning Research.

Judd, D. R. (1983), 'From Cowtown to Sunbelt City: Bootsterism and Economic Growth in Denver' in Fainstein *et al.* (eds), pp.167–201.

Kain, J. (1968), 'Housing Segregation, Negro Employment and Metropolitan Decentralisation', *The Quarterly Journal of Economics*, 82, pp.175–97.

Katznelson, I. (1981), *City Trenches: Urban Politics and the Patterning of Class in the United States*, New York: Pantheon Books.

Kelley, J., McAllister, I. and Mughan, A. (1985), 'The Decline of Class Revisited: Class and Party in England, 1964–1979', *American Political Science Review*, 79.(3), pp.719–37.

Kenyon, P. (1989), 'People and Property Development', in Business in the Community (1989), p.2.

Khanum, S. (1991), 'War Talk: What do British Muslims Think About the Conflict in the Gulf', *New Statesman Society*, 87 (1460), 1 February, pp.12–13.

King, D. (1987a), *The New Right: Politics, Markets and Citizenship*, Basingstoke: Macmillan.

King, D.S. (1987b). 'The State, Capital and Urban Change in Britain', in Feagin and Smith, (eds), 1987.

King, D.S. (1989), 'Voluntary and State Provision of Welfare as Part of the Public-Private Continuum: Modelling the 'Shifting Involvements' in Britain and the USA' in Ware (ed.) (1989), pp.29–54.

King, D.S. (1990), 'Economic Activity and the Challenge to Local Government', in King and Pierre (eds) (1990), pp.265–87.

King, D.S. and Pierre, J. (eds) (1990), *Challenges to Local Government*, London, Newbury Park and New Delhi: Sage.

King, R. (ed.) (1983), *Capital and Politics*, London: Routledge & Keagan Paul.

Kirk, R. (1953), *The Conservative Mind*, Chicago: Regnery.

Kirkby, R.J.R. (1988), *Urbanisation in China: Town and Country in a Developing Economy 1949-2000 AD*, London: Routledge.

Kotz, N. (1982), 'Corporate Philanthropy: The Issues Today' in Council on Foundations (1982), pp.14–16.

Kristol, I. (1983), *Reflections of a Neo-Conservative*, New York: Basic Books.

Labour Party (1989), *Meet the Challenge, Make the Change: A New Agenda for Britain (Final Report of Labour's Policy Review for the 1990s)*, London: Labour Party.

Labour Party (1991), *Opportunity Britain: Labour's Better Way for the 1990s*, London: Labour Party.

Laconte, P. (1977) 'Effects and Dangers of Extreme Urbanisation: The High-density–Low-Rise Layout as a Possible Planning Remedy' in Polunin (ed.) (1980) pp.235–59

Lampe, D. (ed.) (1988), *The Massachusetts Miracle: High Technology and Economic Revitalization*, Cambridge, Mass: Massachusetts Institute of Technology Press.

Lascelles, D. (1991), 'Costly Lure of USA Banking', *Financial Times*, 31 January.

274

REFERENCES

Lawless, P. (1981), *Britain's Inner Cities: Problems and Policies*, London: Harper and Row.

Le Gales, P. (1990), 'Local Economic Policies, the State and City Councils: Coventry and Rennes', in King, and Pierre (eds) (1990), pp.122–44.

Le Lohe, M.J. (1989), 'The Performance of Asian and Black Candidates in the British General Election of 1987', *New Community*, 15 (2), pp.159–70.

Leadbeater, C. (1988), 'Power to the Person', *Marxism Today*, October, pp.14–19.

Levačić, R. (1987), *Economic Policy Making: Its Theory and Practice*, Brighton: Wheatsheaf.

Levinson, M. (1991), 'New Blueprints for the Left: Seeking (again) An Economic Pitch for Democrats', *Newsweek*, 1 April, p.25.

Levitas, R. (ed.) (1986a), *The Ideology of the New Right*, Cambridge: Polity Press.

Levitas, R. (1986b), 'Ideology and the New Right', in Levitas, (ed.) (1986a).

Levitas, R. (1986c), 'Competition and Compliance: The Utopias of the New Right', in Levitas, (ed.) (1986a).

Lord, S. (1989), 'Lesson From America' in Business in the Community (1989), p.3.

Loughlin, M., Gelfand, M.D. and Young, K. (eds) (1985), *Half a Century of Municipal Decline, 1935–1985*, London: Allen and Unwin.

Lovrich, N. and Neiman, M. (1984), *Public Choice Theory in Public Administration: An Annotated Bibliography*, New York: Garland.

Lublin, J.S. (1986), 'USA Poverty Rate Slipped to 14 Per Cent in '85, but Number of Poor Still Relatively High', *The Wall Street Journal*, 27 August.

M. (1984), *Political Parties and Black People: Participation, Representation and Exploitation*, London: The Runnymede Trust.

Mandič, S. (1990), 'Housing Provision in Yugoslavia: Changing Roles of the State, Market and Informal Sectors', in van Vliet and van Weesep (eds) (1990), pp.259–72.

Manley, J. F. (1983), 'Neo-pluralism: A Class Analysis of Pluralism 1 and Pluralism 2', *American Political Science Review*, 77, (2), pp.368–83.

Marable, M. (1985), *Black American Politics: From the Washington Marches to Jesse Jackson*, London: Verso.

Martz, L., Thomas, R., Friday, C., McCormick, J., Carroll, G., Murr, A. and Katel, P. (1990), 'Burning Up Billions', *Newsweek*, 21 May, pp.34–8.

Mason, S. (1984), *Flowers in a Field of Thorns: Volume III*, Cambridge Mass.: Jefferson Park Writing Center.

Matney, W.C. and Johnson, D.L. (1983), *America's Black Population 1970 to 1982: A Statistical View*, Washington DC: US Department of Commerce, Bureau of the Census.

Mayer, V. (1991), 'Community and Regional Development Slashed 13.2 Per Cent', *Nation's Cities Weekly*, 14 (6) p.7.

McCarthy, M. (1986), *Campaigning for the Poor: CPAG and the Politics of Welfare*, London: Croom Helm.

McCarty, K.S. (1991), 'How are Cities and Towns Coping with Gulf Crisis?', *Nation's Cities Weekly*, 14. (4), p.1.

McKay, D. (1984), Essex Papers in Politics and Government 16, *A Reappraisal of Public Choice Theory of Intergovernmental Relations*, University of Essex, Department of Government.

McLean, I. (1987), *Public Choice: An Introduction*, Oxford: Blackwell.

McRobie, G. (1990), *Small is Possible*, 2nd edn, London: Abacus.

Meadows, D.H., Meadows, D.L., Randers, J. and Behrens III, William W. (1972),

REFERENCES

The Limits to Growth: A Report for the Club of Rome's Project on the Predica-ment of Mankind, New York: Universe Books.

Medical Foundation (1984), *General Information Report Including The Medical Foundation Program Evaluation Guidelines*, Boston, Mass.: The Medical Foun-dation.

Merkowitz, D. and Cooney, D. (eds) (1986), *The 1986 Guide to Federal Policy and the Region*, Washington DC: Northeast-Midwest Institute.

Messina, A.M. (1987), 'Ethnic Minority Representation and Party Competition in Britain: The Case of Ealing Borough', *Political Studies*, 35, pp.224–38.

Meyer, J.A. (ed.) (1982), *Meeting Human Needs: Toward a New Public Philosophy*, Washington DC: American Enterprise Institute.

Mieszkowski, P. (1979), *Studies of Prejudice and Discrimination in Urban Housing Markets*, Boston: Federal Reserve Bank of Boston.

Miller, D. and Estrin, S. (1986), 'Market Socialism: A Policy for Socialists', in Forbes, (ed.).

Miller, N.R. (1983), 'Pluralism and Social Choice', *American Political Science Review*, 77(3), pp.734–47.

Mills,L. and Young, K. (1985), *Local Authority Capacity for Economic Develop-ment*, for ESRC Inner Cities Programme, London: Policy Studies Institute.

Moir, C. (1990), 'Bankers Ponder the Return of Prudence', *Financial Times* , 23 November.

Moore, C. and Richardson, J.J. with Moon, J. (1989), *Local Partnership and the Unemployment Crisis in Britain*, London: Unwin Hyman.

Mortished, C. (1991), 'Olympia and York Downgraded', *Estates Gazette*, 18 May, pp.70–1.

Muller, E.N. and Opp, K. (1986), 'Rational Choice and Rebellious Collective Action', *American Political Science Review*, 80(2), pp.471–87.

Mullins, P. (1991), 'The Identification of Social Forces in Development as a General Problem in Sociology: A Comment on Phal's Remarks on Class and Consump-tion Relations as Forces in Urban and Regional Development', *International Journal of Urban and Regional Research*, 15, pp.119–26.

Murray, C. (1984), *Losing Ground*, New York: Basic Books.

Murray, C. (1989), 'Underclass: the Alienated Poor are Devastating America's Cities. Is the Same Happening Here?' *The Sunday Times Magazine*, 26 Novem-ber, pp.26–46.

Murray, R. (1985), 'Benetton Britain: The New Economic Order', reprinted in Hall and Jacques (eds) (1989a), pp.54–64.

Murray, R. (1988), 'Life After Henry Ford', *Marxism Today*, October.

National Council for Civil Liberties (NCCL) (1980), *Southall, 23 April 1979: The Report of the Unofficial Committee of Enquiry*, London: NCCL.

National Council for Voluntary Organisations (NCVO) (1985), *Management Short Courses in Cooperation with Local Organisations, 1985–86*, London: NCVO.

Neil, A. (1982), 'America's Latin Beat: A Survey of South Florida', *The Economist*, Supplement, 16 October.

New York Chamber of Commerce and Industry (1983), 'The Business Marketing Department of the New York Chamber of Commerce and Industry', Infor-mation Paper, New York City: Chamber of Commerce and Industry.

New York, City of, (1983), *Invest in Your Future and the Future of An Entire City*, New York City.

The Newcastle Initiative (1990), *The Newcastle Initiative*, Newcastle Upon Tyne: The Newcastle Initiative with the assistance of the CBI and Urban Regeneration.

REFERENCES

Newton, K. (1976), *Second City Politics: Democratic Processes and Decision-making in Birmingham*, Oxford: Clarendon Press.

North, D. (1986), 'A Neo-classical Theory of the State', in Elster, (ed.), pp.248–60.

North Staffordshire Chamber of Commerce and Industry (1985), 'Letter to Mr. Norman Lamont, M.P., Minister for Industry', *Focus*, 35(1), pp.4–5.

Norton, A. with the assistance of Mills, L. (1990), 'The Role of the British Local Government Chief Executive: A Contribution Towards the International Comparison of Political and Staff Elements and their Relationships Within Local Authority Constitutions', Paper for the ECPR Joint Session Workshops, Bochum, Germany, Birmingham: Institute of Local Government Studies.

Nozick, R. (1974), *Anarchy, State and Utopia*, London: Blackwell.

Office of Population Censuses and Surveys (OCPS) (1986), *Population Trends, 46*, Winter, London: OCPS.

Olsen, J.P. (1976), 'Organisational Participation in Government', unpublished paper, University of Bergen.

Olson, M. (1965), *The Logic of Collective Action*, Cambridge Mass.: Harvard University Press.

Oppenheim, C. (1990), *Poverty: The Facts*, London: Child Poverty Action Group.

Organisation for Economic Cooperation and Development (OECD) (1987), *Recent Trends in International Direct Investment*, Paris: OECD.

Organisation for Economic Cooperation and Development (OECD), (1988), *Transport and the Environment*, Paris: OECD.

Osborne, D. (1989) 'They Can't Stop Us Now', *The Washington Post Magazine*, 30 July.

Ostrom, E. and Whitaker, G. (1974), 'Community Control and Governmental Responsiveness: The Case of Police in Black Neighbourhoods', in Rogers, and Hawley, (eds), pp.303–34.

Ostrom, V. (1976), 'The American Experiment in Constitutional Choice', *Public Choice*, 27, Fall, pp.1–12.

Overbeek, H. (1990), *Global Capitalism and National Decline: The Thatcher Decade in Perspective*, London: Unwin Hyman.

Øyen, E. (ed.) (1990), *Comparative Methodology: Theory and Practice in International Social Research*, Sage Studies in International Sociology 40, sponsored by the International Sociological Association, Newbury Park, London, New Delhi: Sage.

Page, C. (1990), 'Political Pendulum Swinging Toward Government Activism', *Nation's Cities Weekly*, 13 (48), p.12.

Page, E.C. and Goldsmith, M.J. (eds) (1987), *Central and Local Government Relations: A Comparative Study of West European Unitary States*, Newbury Park, London, New Delhi: Sage.

Palmer, J., Sweeding, T. and Torrey, B. (eds) (1988), *The Vulnerable*, Washington DC: The Urban Institute.

Palmer, J.L. and Sawhill, I.V. (eds) (1984), *The Reagan Record: An Assessment of America's Changing Domestic Priorities*, Cambridge Mass: Ballinger Publishing Company for the Urban Institute.

Peirce, N.R. (1991), 'Bush Turnback Plan Sounds Nice but it's Irrelevant', *Nation's Cities Weekly*, 14(7), p.12.

Perkins, E. (1975), *Home is a Dirty Street: The Social Oppression of Black Children*, Chicago: Third World Press.

Perkins, J. (ed.) (1987), *A Conservative Agenda for Black Americans*, Washington DC: The Heritage Foundation.

REFERENCES

Perry, H. L. (1990a), 'Recent Advances in Black Electoral Politics', *Political Science and Politics*, xxiii(2), p.141.

Perry, H.L. (1990b), 'The Evolution and Impact of Biracial Coalitions and Black Mayors in Birmingham and New Orleans' in Browning R. P. *et al.* (eds) (1990).

Peterson, G.E. and Lewis, C.W. (eds) (1986), *Reagan and the Cities*, Washington DC: The Urban Institute Press.

Peterson, G.E. (1984), 'Federalism and the States: An Experiment in Decentralisation', in Palmer and Sawhill (eds), pp.217–59

Peterson, G.E. (1986), 'Urban Policy and the Cyclical Behaviour of Cities' in Peterson and Lewis (eds) (1986), pp.11–35

Peterson, P. (1981), *City Limits*, Chicago: University of Chicago Press.

Phal, R.E. (1989), 'Is the Emperor Naked? Some Questions on the Adequacy of Sociological Theory in Urban and Regional Research', *International Journal of Urban and Regional Research*, 13, pp.709–20.

Phillips, D.R. and Yeh, A.G.O. (1987), *New Towns in East and South-East Asia: Planning and Development*, Oxford: Oxford University Press.

Pilkington, Sir A. (1984), *Business in the Community: Where do we go from Here?*, special reprint of article in April 1984 edition of *Policy Studies*, vol.4 pt.4, (distributed by BITC).

Pinch, S. (1985), *Cities and Services: The Geography of Collective Consumption*, London, Routledge and Kegan Paul.

Piore, M. and Sable, C. (1984), *The Second Industrial Divide: Possibilities for Prosperity*, New York: Basic Books.

Podhoretz, N. (1979), *Breaking Ranks* New York: Harper and Row.

Polunin, N. (ed.) (1980), *Growth Without Ecodisasters? Proceedings of the Second International Conference on Environmental Future (2nd ICEF) Held in Reykjavik, Iceland, 5–11 June 1977*, London: Macmillan.

Porritt, J. and Winner, D. (1989), *The Coming of the Greens*, London: Fontana.

Porritt, J. (1990), *Where on Earth are we Going?*, London: BBC Books.

President's Commission on National Goals (1960), *Goals for Americans: Programs for Action in the Sixties*, New York: American Assembly.

Preston, M.B., Lenneal J.H. and Puryear, P. (eds) (1982), *The New Black Politics: The Search for Political Power*, New York: Longman.

Purkis, A. (1985), *Voluntary Organisations and Government: Reflections from Reagan's America*, London: National Council for Voluntary Organisations.

Redding, City of (1990), *Foundations for the Future: A Report on the State of the City Redding, California, January 1990*, Redding: City of Redding.

Rees, G. and Lambert, J. (1985), *Cities in Crisis: The Political Economy of Urban Development in Post-War Britain*, London: Edward Arnold.

Reich, R. (1991), *The Work of Nations: Preparing Ourselves for 21st. Century Capitalism*, New York: Knopf.

Reid, T. (1990), *Guide to European Community Grants and Loans (UK Edition)*, Newbury, Berkshire: Eurofi plc.

Rex, J. and Tomlinson, S. (1979), *Colonial Immigrants in a British City: A Class Analysis*, London: Routledge.

Rhodes, R. (1986), *The National World of Local Government*, London: Allen and Unwin.

Rhodes, R. (1991), 'ESRC Research Initiative on Local Government', Paper presented at the Political Studies Association (UK) Annual Conference, University of Lancaster.

Rich, M.J. (1986), 'Learning to Live With Less: Federal Aid for Housing and

REFERENCES

Community Development in the 1980s', *American Political Science Association Urban Politics and Urban Policy Newsletter*, 1(1), pp.11–21.

Rich, S. (1986), 'Average Family's Income Up Little Over Eleven Years: Well-to-Do, Older People Make Most Gains', *The Washington Post*, 25 August.

Riddell, P. (1990a), 'The Bottom Line for Bush', *Financial Times*, 8 June.

Riddell, P. (1990b), 'Washington Mayor Jailed Ten Days Before City Polls', *Financial Times*, 27 October.

Rinehart, D.G. (1986), 'How One City Tests Pulse of Community: Focus Groups Use Business Technique to Discover Glitches', *Nation's Cities Weekly*, vol.9, 10 November.

Robinson, P. (1990), 'The High-Octane Semiconductor', *Byte*, January, pp.251–8.

Rogers, D. and Hawley, W. (eds) (1974), *Improving the Quality of Urban Management*, Beverley Hills: Sage.

Rosenthal, U., Charles, M.T. and 't Hart, P. (1989), *Coping With Crises: The Management of Disasters, Riots and Terrorism*, Springfield Ill.: Charles C. Thomas.

Rothbard, M. (1978), *For A New Liberty: The Libertarian Manifesto*, 2nd edn, London and New York: Collier-Macmillan.

Ruckelshaus, W.D. (1982), 'Business Considerations In Corporate Philanthropy' in Council on Foundations (1982) pp.30–1.

Ryan, B. (1990), 'Farewell to Chips?', *Byte*, January, pp.237–49.

Salamon, L.M. and Abramson, A.J. (1984), 'Governance: The Politics of Retrenchment' in Palmer and Sawhill (eds), pp.31–68.

Salamon, L.M, and Abramson, A.J. (1990) *The Federal Budget and the Non-profit Sector: FY 1991 ; A Report to the Independent Sector*, Washington DC: for the Independent Sector.

Sale, K. (1980), *Human Scale*, London: Secker and Warburg.

Saunders, P. (1979), *Urban Politics: A Sociological Interpretation* London: Hutchinson.

Savage, P.S. and Robins, L. (eds) (1990), *Public Policy Under Thatcher*, Basingstoke: Macmillan.

Sawhill, I.V. and Stone, C.F. (1984), 'The Economy: The Key to Success' in Palmer and Sawhill (eds) (1984), pp.69–105

Scarman, Lord Chief Justice, (1981), *The Brixton Disorders 10–12 April 1981: Report of an Inquiry by the Rt. Hon. Lord Scarman OBE*, London: HMSO.

Schumacher, E.F. (1974), *Small is Beautiful: A Study of Economics as if People Mattered*, 2nd edn, London: Sphere.

Scruton, R. (1984), *The Meaning of Conservatism*, 2nd edn, London: Macmillan.

Seale, B. (1970), *Seize the Time: The Story of the Black Panther Party*, London: Arrow and Hutchison.

Sert, J.L. (1942), *Can Our Cities Survive?* Cambridge, Mass.: Harvard University Press.

Shafroth, F. (1989), 'The Reagan Years and the Nation's Cities', *The Municipal Year Book 1989*, Washington DC: International City Management Association, pp.115–28.

Shafroth, F. (1991), 'The Bush Plan to Overhaul the Banking System', *Nation's Cities Weekly*, 14. (6), p.16.

Sheffield Development Corporation (SDC) (1989), *Meeting the Challenge*, Sheffield: SDC.

Sheffield, J.F. and Hadley, C.D. (1984), 'Racial Voting in a Biracial City: A Reexamination of Some Hypotheses', *American Politics Quarterly*, 12(4), pp.449–64.

REFERENCES

Sills, A., Taylor, G. and Golding, P. (1988), *The Politics of the Urban Crisis*, London: Unwin Hyman.

Simpson, R.P. and Jung, C.S. (1990), *California Counties on the Fiscal Fault Line: A Study of the Financial Condition of Californian Counties*, Sacramento, Calif. California Counties Foundation.

Smerk, G.M. (1991), *The Federal Role in Urban Mass Transportation*, Bloomington and Indianapolis: Indiana University Press.

Smith, A. (1991), 'Birmingham Financial Centre: Weathering the Storm', *Investor's Chronicle*, 24 May. pp.53–8.

Smith, M.P. (1979), *The City and Social Theory*, New York: St. Martin's Press.

Sowell, T. (1975), *Race and Economics*, New York: David McKay Co.

Spring, W.J. (1987), 'Youth Unemployment and the Transition from School to Work: Programs in Boston, Frankfurt and London', *New England Economic Review*, March–April, pp.3–16.

Sternlieb, G. and Burchett, R.W. (1973), *Residential Abandonment*, New Brunswick: Rutgers University Press.

Stewart, J. and Stoker, G. (eds) (1989), *The Future of Local Government*, Basingstoke: Macmillan.

Stockman, D.A. (1986), *The Triumph of Politics: The Crisis in American Government and How it Affects the World*, London: Coronet.

Stoke-on-Trent City Council, Staffordshire County Council, North Staffordshire Chamber of Commerce and Industry, North Staffordshire Trades Council and the British Ceramic Manufacturers Federation (1985), *A Case for Inner Area Status for Stoke-on-Trent*, Stoke-on-Trent: Stoke-on-Trent City Council.

Stoker, G. (1988), *The Politics of Local Government*, Basingstoke: Macmillan.

Stoker, G. (1990), 'Regulation Theory, Local Government and the Transition from Fordism', in King and Pierre (eds) (1990), pp.242–64.

Stone, C. (1976), *Economic Growth and Neighborhood Discontent*, Chapel Hill, NC: University of North Carolina Press.

Stone, C.N. (1986), 'Partnership New South Style: Central Atlanta Progress', in Davis, P. (ed.) (1986), pp. 100–10.

Stone, C.N. (1989), *Regime Politics: Governing Atlanta*, Lawrence Kan.: University of Kansas Press.

Summers, M. and Klinker, P. (1990), 'The Election of John Daniels as Mayor of New Haven', *Political Science and Politics*, xxiii(2), pp.142–5.

Swanstrom, T. (1988), 'Semisoverign Cities: The Politics of Urban Development', *Polity*, 21 (1), pp.83–110.

Szelenyi, I. (1983), *Urban Inequalities Under State Socialism*, Oxford: Oxford University Press.

TAFT Corporate Information System (1987), *TAFT Corporate Giving Directory: Comprehensive Profiles of America's Major Corporate Foundations and Charitable Giving Programs*, 8th end, TAFT.

Taaffe, P. and Mulhearn, T. (1988), *Liverpool: A City that Dared to Fight*, London: Fortress.

Thompson, G. (ed.) (1989), *Industrial Policy: US and UK Debates*, Economy and Society Series, London: Routledge.

Thompson, P.J. (1990), 'David Dinkins' Victory in New York City: The Decline of the Democratic Party Organisation and the Strengthening of Black Politics', *Political Science and Politics*, xxiii(2), pp.145–8.

Thrift, N. and Forbes, D. (1986), *The Price of War: Urbanisation in Vietnam 1954–85*, London: Allen and Unwin.

Tiebout, C.M. (1956), 'A Pure Theory of Local Expenditures', *Journal of Political Economy*, 64, pp.416–24.

Tolley (1986), *Tolley's Charities Manual*, Croydon: Tolley Publishing Company.

Turner, S. (1990), 'Hispanic Americans', *Collier's International Year Book 1990*, London and New York: Collier Macmillan, pp.72–81.

United Community Planning Corporation (UCPC) (1983), *Annual Report*, Boston, Mass.: UCPC.

United Way of America (1983), *Annual Report*, Washington DC: United Way of America.

United Way of Massachusetts Bay (UWMB) (1981), *Annual Report*, Boston, Mass.: UWMB.

US Government Printing Office (1988), *Abstract of Statistics*, Washington DC: US Government Printing Office.

Verity Jr, W.C. (1982), 'The Role of Business in Community Affairs' in Council on Foundations (1982) pp.32–34.

Visiting Nurse Association of Boston (VNAB) (1982a), *Annual Report: Meeting the Challenge*, Boston, Mass.: VNAB.

Visiting Nurse Association of Boston (VNAB) (1982b), *House Call*, Spring, Boston, Mass.: VNAB.

Visiting Nurse Association of Boston (VNAB) (1982c), *Management Group Manual*, Boston, Mass.: VNAB.

Vize, R. (1991a), 'Challenge Dilemma for Capped Councils', *Public Money*, 4 April, p.1.

Vize, R. (1991b), 'Councils Fire the First Salvo: The English Local Authority Associations are Making Their Opening Gambit in the 1992–93 Spending Round', *Public Money*, 27 June, p.8.

van Vliet, W. and van Weesep, J. (eds) (1990), *Government and Housing: Developments in Seven Countries*, Urban Affairs Annual Reviews, Vol. 36, Newbury Park, London, New Delhi: Sage.

Wakeford, R. (1990), *American Development Control: Parallels and paradoxes from an English Perspective*, London: HMSO.

Walker, D. and the Bay Area Study Group, (1990), 'The Playground of USA Capitalism? The Political Economy of the San Francisco Bay Area in the 1980s', in Davis, *et al.*, pp.3–82.

Ware, Alan (ed.) (1989), *Charities and Government*, Manchester: Manchester University Press.

Waste, R.J. (1990), 'What is Urban Political Economy?', *The American Political Science Association Urban Politics and Urban Policy Section Newsletter*, 4(3) Winter, pp.6–10.

Weber, M.P. (1990), 'Pittsburgh: Rebuilding A City' in Bernard (ed.) (1990), pp.227–45.

Weinberg, Sir M. (1990), 'Companies That Do Care', *First*, pp.8–10.

Welch, S. and Studlar, D.T. (1983), 'The Policy Opinions of British Political Activists', *Political Studies*, 31, pp.604–19.

Whiteley, P. and Winyard, S. (1984), 'The Origins of the New Poverty Lobby', *Political Studies*, 32, pp.32–54.

Wildavsky, A. (1987), 'Choosing Preferences By Constructing Institutions: A Cultural Theory of Preference Formation', *American Political Science Review*, 81(1), pp.3–21.

Williams, E.N. (1982), 'Black Political Progress in the 1970s: The Electoral Arena', in Preston *et al.* (eds), pp.73–108.

REFERENCES

Williams, J. (1987), *Eyes on the Prize: America's Civil Rights Years 1954–1965*, New York: Viking Penguin.

Williams, K., Cutler, T., Williams, J. and Haslam, C. (1989), 'The End of Mass Production?' in Thompson (ed.) (1989).

Williams, R. (1976), *Keywords: A Vocabulary of Culture and Society*, New York: Oxford University Press.

Wilson, D.C. (1989), 'New Trends in the Funding of Charities: The Tripartite System of Funding' in Ware (ed.) (1989), pp.54–81.

Wiseman, J. (1989), *Cost, Choice and Political Economy*, Aldershot: Edward Elgar Publishing.

Wollack, L. (1991), 'White House Hears Local CDBG Concerns, Sununu: Make Us an Offer We Can't Refuse', *Nation's Cities Weekly*, 14 (7), p.1.

Wolman, H. (1990), 'The Reagan Urban Policy and its Impacts', in King and Pierre (eds), pp.145–72.

Wright, J.R. and Goldberg, A.S. (1985), 'Risk and Uncertainty as Factors in the Durability of Political Coalitions', *American Political Science Review*, 79(3), pp.704–18.

INDEX

INDEX

Choate, P. 114
choice: rational 22, 23, 36; residential 42; *see also* public choice
Christopher, W. 191
cities: central 53, 60, 61, 63; distress in 99–103; inner 143–4, 147, 188, 208; largest 61, 76, 92; Latino 173; opposition in 136; satellite 63; strong-mayor and weak-mayor 112; sub- 15, 63
Citizen's Charter 138–9
City Action Teams 149
city cultures 2
city trenches 157–9
city government 8, 111–13
civic expectation 198
civil liberties 152
civil rights 14, 171, 172, 174, 178
Clark, C. 239
Clark, G. 92, 153, 154, 155
Clark, P. 188
Clark, T. N. 2, 42, 45
Clarke, S. 33, 46
class *see* social class
classlessness 8, 221–2
Cleveland (UK) 251
Cleveland (US) 76
Club of Rome 259
Cold War 127
collective action 157, 204, 205, 239, 257
collectives 171, 254
collectivism 19, 65
Colorado 64
Columbus (Ohio) 250
commercialisation 229, 236–7
Commission for Racial Equality (CRE) 180, 188
Common Agricultural Policy (CAP) 150
communications networks 247
communism *see* Eastern Europe
community architecture 251, 252, 263
Community Charge *see* Poll Tax
community development 237; corporations 44, 248; *see also under* development
Community Development Project 100
Community Relations Council 189
Community Service Volunteers 180
comparative economic indicators 51
competition 24, 65, 157, 238; foreign 8, 56, 67, 85, 95, 96, 99
competitiveness 53; financial sector 99;

future 68; international 15, 56, 85–6, 95
compulsory competitive tendering 132
computer industry 24, 67, 68, 93, 106; 'clone' systems 69
conflict: class 5; cooperation, and dissent 184–9; ethnic and racial groups 5, 159, 172, 193; ideological 129; interest groups 195; partnership and 39, 257–9; place and content of 158; political 31, 45, 132, 142, 157, 210; rules of 158; social 7, 170
congestion 255, 257
Congressional Joint Economic Committee 79
Congressional Quarterly 114
Connecticut 57, 109
consent 168–9
conservationist issues *see* green issues
conservatism 106, 152
Conservative Agenda for Black Americans, A 179
Conservative government 44–6, 162; and employment in depressed regions 94; and European cooperation 86; and local government 132, 143, 144; and market forces 247
Conservative party 82, 139, 140, 183; green orientation 257; Heseltine and 7, 13; new right influence on 21
'Conservative Revolution' 11–12
conservatives 26, 79, 179; black 171, 178; business leadership/community goals 17; councils 136; principles 12; programmes 10; public housing sales 249; strength of political forces 177; traditional 20; view of citizens' rights 150
constraints: expenditure 210; institutional 10; local government 170, 213–14; market, upon government 38; political 10–12, 126–8
consumer goods 33, 54
consumption 24, 29, 30, 31, 34, 256
contributions 9, 228, 231
controls: community 27, 249, 253; budget 109; capital spending 136, 213; police 168; production 66; resources 163; social 261; worker 254
Cooke, P. 100
cooperation 16; close association 167–8; conflict and dissent 184–9

286

Mason, Sylvia 238
mass production 33, 47, 65
Massachusetts 57, 67, 109, 203; Bay
  Community 232; Commonwealth of
  40, 44, 198, 200, 201, 238;
  Department of Mental Health 233,
  235; Department of Social Services
  233; high technology clustering 66;
  innovative programmes 44; personal
  income 57; Task Force on Private
  Sector Initiatives (GTF) 197–201; see
  also Boston
mathematical economics 22
Mayer, V. 21, 38, 46
Meadows, D. H. 259, 260
Medicaid 76, 223
Medical Foundation, The 234
Medicare 117, 223, 224
'Mega' shopping centres 255
'megaclusters' 63
Merseyside see Liverpool
Messina, A. M. 151
metropolitan areas/districts 58, 130,
  136; largest twenty 59
Mexican-Americans 173, 243
MFP (Multi Function Polis) 2
Miami 74, 186
Michigan 53, 57
microtechnology 247
middle class 61, 221; women 105
Middle East 127
Mieszkowski, P. 75
Militant Tendency 45, 137, 142, 148
militants 184, 185–6, 189; council
  activism 138; organisations 8, 188;
  protests 243
Miller, D. 18, 160
Mills, L. 40, 144
Minneapolis 63
minorities see ethnic minorities
mobility 261
Moir, C. 85
Mon Valley 251, 252
Mondale, Walter 11
monetary policies 50
monitoring 136, 138, 139, 146, 168, 250;
  see also evaluation
Moore, C. 46, 210
moral high-ground 176
mortgages 71
Mortished, C. 97
Moscow 261
motivations 163, 167

motor car 255–6
Moynihan, D. 113
Mulhearn, T. 45
Muller, E. N. 161
Mullins, P. 31
multi-ethnic coalitions 172, 173, 174–5,
  184
multinational enterprises 55–6, 93
Murray, C. 104–5, 106, 238
Murray, R. 34, 35, 40, 47
MXDs (mixed-use developments) 63

National Association for the
  Advancement of Coloured People
  184
National Association of Freedom 21
National Congress for Puerto Rican
  Rights 173
National Council for Civil Liberties
  (NCCL) 181, 186
National Council on Philanthropy
  (NCOP) 217
National Council for Voluntary
  Organisations (NCVO) 228
National Guard 6
National Health Service 156
National Puerto Rican Coalition 173
National Savings Certificates 229
National Westminster Bank 73, 74
nationalism 177, 178, 185
Neil, A. 186
Neiman, M. 22–3, 25–6, 28
neo-conservatives 20
neo-corporatist approaches 3
neo-liberals 20
neo-Marxism 15, 28, 29–31
networks 105, 164–5, 245, 247, 248, 255
neutrality 181, 182
New Commonwealth 103
New England 73, 74
New Federalism 116, 128
New Haven 175–6
New Jersey 57
new realism 142
'new right' 4, 8, 9, 14, 19–22
'new urban left' 129, 135, 143
New York 58, 109, 112, 213, 248, 260;
  black electoral success 175, 176;
  black population 172; corruption
  scandal 176; financial services 53, 56,
  97, 98; Jewish voters angered 177;
  racism 191, 192
Newark 6

Newcastle upon Tyne 91, 211
Newton, K. 160
Nixon, Richard M. 114
noise disturbance 256
non-metropolitan districts 130
non-profit organisations 167, 213,
  216–20, 235, 238, 241;
  commercialisation of 229;
  dependency of 245; expansion 17;
  expected to contribute more 9;
  governments and 222–8; role 23, 221;
  support to 143; see also charities;
  foundations; voluntary organisations
North, D. 23
North Sea oil and gas 83
Northeast-Midwest Institute 121
north/south divide 88
Norton, A. 1
Norway 78
Nozick, R. 20
NPC (new political culture) 153–6
NUMMI (New United Motor
  Manufacturing Inc.) 55

O & Y 97–8
obsolete industries 61
OECD (Organisation for Economic
  Cooperation and Development) 55,
  77, 82, 256, 257, 258
office centres and concentrations 63;
  vacancy rates 98
Office of Management and Budget 21
Olsen, J. P. 163, 166
Olson, M. 165
Olympic Games (1996) 252
'Operation Swamp' 187
Opp, K. 161
Oppenheim, C. 52, 104, 105, 106
Oppenheimer, J. A. 161
options exchanges 56
Osborne, 249
Ostrom, E. 26, 27
out-of-town sites 38, 170
Overbeek, H. 7
overseas companies 106
ownership: car 255; common 253; state
  153
Øyen, E. 3

Page, C. 204
Page, E. C. 1
Paisley 242
Palmer, J. L. 77

Palo Alto 253
Pareto principle 27, 36
Paris 57, 248, 260
participation 161, 202, 263; in
  government 162, 163, 167; group
  166; local 25; minority 184; private
  sector 167; problems of 254
partnerships 196, 222, 232; and
  'collectivisation' 204–5; community
  258, 263; conflict and 39, 257–9;
  effective 238; experience and
  knowledge about 243; initiatives 208;
  local 171, 241, 244, 245; models
  200–1; private sector 80; public/
  private 17, 162, 165, 195, 198–9, 201,
  207, 208, 210, 213, 214; shared vision
  17, 209–11; tensions and 210; types
  of 197–201; variety of 203; visionary
  leadership required in 209
peace dividend 127, 224
Peirce, N. R. 129
penalties 135, 136, 137
Penfield, Judge Thomas 190
Penne, R. L. 75
pension funds 98
Per Cent Club, The 239–40
Perkins, E. 178, 179
Perry, H. L. 174–5
Peterson, G. E. 75, 77, 108, 115
Peterson, P. 37, 41–2, 45
Phal, R. E. 31
Philadelphia 194
philanthropy 198, 199, 200, 217, 221,
  228
Phillips, D. R. 2
Phoenix Initiative (Business in the
  Cities) 211
Pierre, J. 38, 39, 48
Pilkington, Sir Alistair 205
Pinch, S. 29
Piore, M. 47, 48
Pittsburgh 194, 251, 252
Planning Aid for Londoners 241
Plant, R. 21, 28
pluralism 18, 31–3, 174
Podhoretz, N. 20
police 27, 185, 186, 187–8, 189;
  brutality 191, 192; clashes with black
  and Asian groups 181; harassment
  180; powers 151, 152; relations with
  ethnic minorities 189; role 168
policy: fiscal 45; formulation 160;
  impact of policy-makers 38–9; issues